METAPHORS WE LIVE BY

METAPHORS
We Live By

GE**...**NSON

The Un**...**cago Press

Chicago and London

94 W to 252 to 610

on 610 take 1st exit after
it crosses the river onto
E. River Rd.

Left at bottom of exit (5 miles
to go from here)

R. on Riesling Blvd. (by
Franks Nursery & 2 car
dealerships)

R on 117th Lane (1st st.)

Apt. Bldg. (Summit Oaks)

The University of Chicago Press, Chicago 60637
The University of Chicago Press, Ltd., London

© 1980 by The University of Chicago
All rights reserved. Published 1980
Printed in the United States of America
94 93 92 91 90 89 88 87 86   9 8 7 6

Library of Congress Cataloging in Publication Data

Lakoff, George.
Metaphors we live by.

Bibliography: p.
1. Languages—Philosophy.   2. Metaphor.
3. Semantics.   4. Truth.   I. Johnson, Mark, joint
author.   II. Title.
P106.L235   401   80-10783
ISBN 0-226-46800-3

Much of the material in all or parts of chapters 1 through 5, 9
through 12, 14, 18, and 21 originally appeared in the article "Conceptual Metaphor in Everyday Language," *Journal of Philosophy*
77, no. 8 (August 1980): 453–86, and is reprinted here with the
kind permission of the editors of the *Journal of Philosophy*.

For Andy and The Gherkin

Contents

Preface   ix

Acknowledgments   xi

1. Concepts We Live By   3

2. The Systematicity of Metaphorical Concepts   7

3. Metaphorical Systematicity: Highlighting and
   Hiding   10

4. Orientational Metaphors   14

5. Metaphor and Cultural Coherence   22

6. Ontological Metaphors   25

7. Personification   33

8. Metonymy   35

9. Challenges to Metaphorical Coherence   41

10. Some Further Examples   46

11. The Partial Nature of Metaphorical
    Structuring   52

12. How Is Our Conceptual System Grounded?   56

13. The Grounding of Structural Metaphors   61

14. Causation: Partly Emergent and Partly
    Metaphorical   69

15. The Coherent Structuring of Experience   77

16. Metaphorical Coherence   87

17. Complex Coherences across Metaphors   97

18. Some Consequences for Theories of
    Conceptual Structure   106

19. Definition and Understanding   115

20. How Metaphor Can Give Meaning to Form   126

21. New Meaning   139

22. The Creation of Similarity   147

23. Metaphor, Truth, and Action   156

24. Truth   159

25. The Myths of Objectivism and Subjectivism   185

26. The Myth of Objectivism in Western
    Philosophy and Linguistics   195

27. How Metaphor Reveals the Limitations of
    the Myth of Objectivism   210

28. Some Inadequacies of the Myth of
    Subjectivism   223

29. The Experientialist Alternative: Giving New
    Meaning to the Old Myths   226

30. Understanding   229

Afterword   239

References   241

Preface

This book grew out of a concern, on both our parts, with how people understand their language and their experience. When we first met, in early January 1979, we found that we shared, also, a sense that the dominant views on *meaning* in Western philosophy and linguistics are inadequate—that "meaning" in these traditions has very little to do with what people find *meaningful* in their lives.

We were brought together by a joint interest in metaphor. Mark had found that most traditional philosophical views permit metaphor little, if any, role in understanding our world and ourselves. George had discovered linguistic evidence showing that metaphor is pervasive in everyday language and thought—evidence that did not fit any contemporary Anglo-American theory of meaning within either linguistics or philosophy. Metaphor has traditionally been viewed in both fields as a matter of peripheral interest. We shared the intuition that it is, instead, a matter of central concern, perhaps the key to giving an adequate account of understanding.

Shortly after we met, we decided to collaborate on what we thought would be a brief paper giving some linguistic evidence to point up shortcomings in recent theories of meaning. Within a week we discovered that certain assumptions of contemporary philosophy and linguistics that have been taken for granted within the Western tradition since the Greeks precluded us from even raising the kind of issues we wanted to address. The problem was not one of extending or patching up some existing theory of meaning but of revising central assumptions in the Western philosophical tradition. In particular, this meant rejecting the possibility of any objective or absolute truth and a host of related assumptions. It also meant supplying an alternative account in which human experience and understanding, rather than objective truth, played the central role. In the process, we have worked out elements of an experientialist approach, not only to issues of language, truth, and understanding but to questions about the meaningfulness of our everyday experience.

Berkeley, California
July 1, 1979

Acknowledgments

Ideas don't come out of thin air. The general ideas in this book represent a synthesis of various intellectual traditions and show the influence of our teachers, colleagues, students, and friends. In addition, many specific ideas have come from discussions with literally hundreds of people. We cannot adequately acknowledge all of the traditions and people to whom we are indebted. All we can do is to list some of them and hope that the rest will know who they are and that we appreciate them. The following are among the sources of our general ideas.

John Robert Ross and Ted Cohen have shaped our ideas about linguistics, philosophy, and life in a great many ways.

Pete Becker and Charlotte Linde have given us an appreciation for the way people create coherence in their lives.

Charles Fillmore's work on frame semantics, Terry Winograd's ideas about knowledge-representation systems, and Roger Schank's conception of scripts provided the basis for George's original conception of linguistic gestalts, which we have generalized to *experiential gestalts*.

Our views about family resemblances, the prototype theory of categorization, and fuzziness in categorization come from Ludwig Wittgenstein, Eleanor Rosch, Lotfi Zadeh, and Joseph Goguen.

Our observations about how a language can reflect the conceptual system of its speakers derive in great part from the work of Edward Sapir, Benjamin Lee Whorf, and others who have worked in that tradition.

Our ideas about the relationship between metaphor and ritual derive from the anthropological tradition of Bronislaw

Malinowski, Claude Lévi-Strauss, Victor Turner, Clifford Geertz, and others.

Our ideas about the way our conceptual system is shaped by our constant successful functioning in the physical and cultural environment come partly from the tradition of research in human development begun by Jean Piaget and partly from the tradition in ecological psychology growing out of the work of J. J. Gibson and James Jenkins, particularly as represented in the work of Robert Shaw, Michael Turvey, and others.

Our views about the nature of the human sciences have been significantly influenced by Paul Ricoeur, Robert McCauley, and the Continental tradition in philosophy.

Sandra McMorris Johnson, James Melchert, Newton and Helen Harrison, and David and Ellie Antin have enabled us to see the common thread in aesthetic experience and other aspects of our experience.

Don Arbitblit has focused our attention on the political and economic implications of our ideas.

Y. C. Chiang has allowed us to see the relationship between bodily experience and modes of viewing oneself and the world.

We also owe a very important debt to those contemporary figures who have worked out in great detail the philosophical ideas we are reacting against. We respect the work of Richard Montague, Saul Kripke, David Lewis, Donald Davidson, and others as important contributions to the traditional Western conceptions of meaning and truth. It is their clarification of these traditional philosophical concepts that has enabled us to see where we diverge from the tradition and where we preserve elements of it.

Our claims rest largely on the evidence of linguistic examples. Many if not most of these have come out of discussions with colleagues, students, and friends. John Robert Ross, in particular, has provided a steady stream of examples via phone calls and postcards. The bulk of the examples in chapters 16 and 17 came from Claudia Brugman, who also gave us invaluable assistance in the preparation of the manuscript. Other examples have come from Don Arbitblit, George Bergman, Dwight Bolinger, Ann Borkin, Matthew Bronson, Clifford Hill, D. K. Houlgate III, Dennis Love, Tom Mandel, John Manley-Buser, Monica Macauley, James D. McCawley, William Nagy, Reza Nilipoor, Geoff Nunberg, Margaret Rader, Michael Reddy, Ron Silliman, Eve Sweetser, Marta Tobey, Karl Zimmer, as well as various students at the University of California, Berkeley, and at the San Francisco Art Institute.

Many of the individual ideas in this work have emerged from informal discussions. We would particularly like to thank Jay Atlas, Paul Bennaceraf, Betsy Brandt, Dick Brooks, Eve Clark, Herb Clark, J. W. Coffman, Alan Dundes, Glenn Erickson, Charles Fillmore, James Geiser, Leanne Hinton, Paul Kay, Les Lamport, David Lewis, George McClure, George Rand, John Searle, Dan Slobin, Steve Tainer, Len Talmy, Elizabeth Warren, and Bob Wilensky.

METAPHORS WE LIVE BY

1

Concepts We Live By

Metaphor is for most people a device of the poetic imagination and the rhetorical flourish—a matter of extraordinary rather than ordinary language. Moreover, metaphor is typically viewed as characteristic of language alone, a matter of words rather than thought or action. For this reason, most people think they can get along perfectly well without metaphor. We have found, on the contrary, that metaphor is pervasive in everyday life, not just in language but in thought and action. Our ordinary conceptual system, in terms of which we both think and act, is fundamentally metaphorical in nature.

The concepts that govern our thought are not just matters of the intellect. They also govern our everyday functioning, down to the most mundane details. Our concepts structure what we perceive, how we get around in the world, and how we relate to other people. Our conceptual system thus plays a central role in defining our everyday realities. If we are right in suggesting that our conceptual system is largely metaphorical, then the way we think, what we experience, and what we do every day is very much a matter of metaphor.

But our conceptual system is not something we are normally aware of. In most of the little things we do every day, we simply think and act more or less automatically along certain lines. Just what these lines are is by no means obvious. One way to find out is by looking at language. Since communication is based on the same conceptual system that we use in thinking and acting, language is an important source of evidence for what that system is like.

Primarily on the basis of linguistic evidence, we have found that most of our ordinary conceptual system is metaphorical in nature. And we have found a way to begin to identify in detail just what the metaphors are that structure how we perceive, how we think, and what we do.

To give some idea of what it could mean for a concept to be metaphorical and for such a concept to structure an everyday activity, let us start with the concept ARGUMENT and the conceptual metaphor ARGUMENT IS WAR. This metaphor is reflected in our everyday language by a wide variety of expressions:

ARGUMENT IS WAR

Your claims are *indefensible*.
He *attacked every weak point* in my argument.
His criticisms were *right on target*.
I *demolished* his argument.
I've never *won* an argument with him.
You disagree? Okay, *shoot!*
If you use that *strategy*, he'll *wipe you out*.
He *shot down* all of my arguments.

It is important to see that we don't just *talk* about arguments in terms of war. We can actually win or lose arguments. We see the person we are arguing with as an opponent. We attack his positions and we defend our own. We gain and lose ground. We plan and use strategies. If we find a position indefensible, we can abandon it and take a new line of attack. Many of the things we *do* in arguing are partially structured by the concept of war. Though there is no physical battle, there is a verbal battle, and the structure of an argument—attack, defense, counterattack, etc.—reflects this. It is in this sense that the ARGUMENT IS WAR metaphor is one that we live by in this culture; it structures the actions we perform in arguing.

Try to imagine a culture where arguments are not viewed in terms of war, where no one wins or loses, where there is no sense of attacking or defending, gaining or losing ground. Imagine a culture where an argument is viewed as a dance, the participants are seen as performers, and the goal is to perform in a balanced and aesthetically pleasing way. In such a culture, people would view arguments differently, experience them differently, carry them out differently, and talk about them differently. But *we* would probably not view them as arguing at all: they would simply be doing something different. It would seem strange even to call what they were doing "arguing." Perhaps the most neutral way of describing this difference between their culture and ours would be to say that we have a discourse form structured in terms of battle and they have one structured in terms of dance.

This is an example of what it means for a metaphorical concept, namely, ARGUMENT IS WAR, to structure (at least in part) what we do and how we understand what we are doing when we argue. *The essence of metaphor is understanding and experiencing one kind of thing in terms of another.* It is not that arguments are a subspecies of war. Arguments and wars are different kinds of things—verbal discourse and armed conflict—and the actions performed are different kinds of actions. But ARGUMENT is partially structured, understood, performed, and talked about in terms of WAR. The concept is metaphorically structured, the activity is metaphorically structured, and, consequently, the language is metaphorically structured.

Moreover, this is the *ordinary* way of having an argument and talking about one. The normal way for us to talk about attacking a position is to use the words "attack a position." Our conventional ways of talking about arguments presuppose a metaphor we are hardly ever conscious of. The metaphor is not merely in the words we use—it is in our very concept of an argument. The language of argument is not poetic, fanciful, or rhetorical; it is literal. We talk about arguments that way because we conceive of them that way—and we act according to the way we conceive of things.

The most important claim we have made so far is that metaphor is not just a matter of language, that is, of mere words. We shall argue that, on the contrary, human *thought processes* are largely metaphorical. This is what we mean when we say that the human conceptual system is metaphorically structured and defined. Metaphors as linguistic expressions are possible precisely because there are metaphors in a person's conceptual system. Therefore, whenever in this book we speak of metaphors, such as AR-GUMENT IS WAR, it should be understood that *metaphor* means *metaphorical concept.*

2

The Systematicity
of Metaphorical Concepts

Arguments usually follow patterns; that is, there are certain things we typically do and do not do in arguing. The fact that we in part conceptualize arguments in terms of battle systematically influences the shape arguments take and the way we talk about what we do in arguing. Because the metaphorical concept is systematic, the language we use to talk about that aspect of the concept is systematic.

We saw in the ARGUMENT IS WAR metaphor that expressions from the vocabulary of war, e.g., *attack a position, indefensible, strategy, new line of attack, win, gain ground,* etc., form a systematic way of talking about the battling aspects of arguing. It is no accident that these expressions mean what they mean when we use them to talk about arguments. A portion of the conceptual network of battle partially characterizes the concept of an argument, and the language follows suit. Since metaphorical expressions in our language are tied to metaphorical concepts in a systematic way, we can use metaphorical linguistic expressions to study the nature of metaphorical concepts and to gain an understanding of the metaphorical nature of our activities.

To get an idea of how metaphorical expressions in everyday language can give us insight into the metaphorical nature of the concepts that structure our everyday activities, let us consider the metaphorical concept TIME IS MONEY as it is reflected in contemporary English.

TIME IS MONEY

You're *wasting* my time.
This gadget will *save* you hours.

7

I don't *have* the time to *give* you.
How do you *spend* your time these days?
That flat tire *cost* me an hour.
I've *invested* a lot of time in her.
I don't *have enough* time to *spare* for that.
You're *running out* of time.
You need to *budget* your time.
Put aside some time for ping pong.
Is that *worth your while?*
Do you *have* much time *left?*
He's living on *borrowed* time.
You don't *use* your time *profitably.*
I *lost* a lot of time when I got sick.
Thank you for your time.

Time in our culture is a valuable commodity. It is a limited resource that we use to accomplish our goals. Because of the way that the concept of work has developed in modern Western culture, where work is typically associated with the time it takes and time is precisely quantified, it has become customary to pay people by the hour, week, or year. In our culture TIME IS MONEY in many ways: telephone message units, hourly wages, hotel room rates, yearly budgets, interest on loans, and paying your debt to society by "serving time." These practices are relatively new in the history of the human race, and by no means do they exist in all cultures. They have arisen in modern industrialized societies and structure our basic everyday activities in a very profound way. Corresponding to the fact that we *act* as if time is a valuable commodity—a limited resource, even money—we *conceive of* time that way. Thus we understand and experience time as the kind of thing that can be spent, wasted, budgeted, invested wisely or poorly, saved, or squandered.

TIME IS MONEY, TIME IS A LIMITED RESOURCE, and TIME IS A VALUABLE COMMODITY are all metaphorical concepts. They are metaphorical since we are using our everyday experiences with money, limited resources, and valuable commodities to conceptualize time. This isn't a necessary way for human beings to conceptualize time; it is tied to our culture. There are cultures where time is none of these things.

The metaphorical concepts TIME IS MONEY, TIME IS A RESOURCE, and TIME IS A VALUABLE COMMODITY form a single system based on subcategorization, since in our society money is a limited resource and limited resources are valuable commodities. These subcategorization relationships characterize entailment relationships between the metaphors. TIME IS MONEY entails that TIME IS A LIMITED RESOURCE, which entails that TIME IS A VALUABLE COMMODITY.

We are adopting the practice of using the most specific metaphorical concept, in this case TIME IS MONEY, to characterize the entire system. Of the expressions listed under the TIME IS MONEY metaphor, some refer specifically to money (*spend, invest, budget, profitably, cost*), others to limited resources (*use, use up, have enough of, run out of*), and still others to valuable commodities (*have, give, lose, thank you for*). This is an example of the way in which metaphorical entailments can characterize a coherent system of metaphorical concepts and a corresponding coherent system of metaphorical expressions for those concepts.

3

Metaphorical Systematicity: Highlighting and Hiding

The very systematicity that allows us to comprehend one aspect of a concept in terms of another (e.g., comprehending an aspect of arguing in terms of battle) will necessarily hide other aspects of the concept. In allowing us to focus on one aspect of a concept (e.g., the battling aspects of arguing), a metaphorical concept can keep us from focusing on other aspects of the concept that are inconsistent with that metaphor. For example, in the midst of a heated argument, when we are intent on attacking our opponent's position and defending our own, we may lose sight of the cooperative aspects of arguing. Someone who is arguing with you can be viewed as giving you his time, a valuable commodity, in an effort at mutual understanding. But when we are preoccupied with the battle aspects, we often lose sight of the cooperative aspects.

A far more subtle case of how a metaphorical concept can hide an aspect of our experience can be seen in what Michael Reddy has called the "conduit metaphor." Reddy observes that our language about language is structured roughly by the following complex metaphor:

IDEAS (or MEANINGS) ARE OBJECTS.
LINGUISTIC EXPRESSIONS ARE CONTAINERS.
COMMUNICATION IS SENDING.

The speaker puts ideas (objects) into words (containers) and sends them (along a conduit) to a hearer who takes the idea/objects out of the word/containers. Reddy documents this with more than a hundred types of expressions in English, which he estimates account for at least 70 percent of the expressions we use for talking about language. Here are some examples:

The CONDUIT Metaphor

It's hard to *get* that idea *across to* him.
I *gave* you that idea.
Your reasons *came through* to us.
It's difficult to *put* my ideas *into* words.
When you *have* a good idea, try to *capture* it immediately *in* words.
Try to *pack* more thought *into* fewer words.
You can't simply *stuff* ideas *into* a sentence any old way.
The meaning is right there *in* the words.
Don't *force* your meanings *into* the wrong words.
His words *carry* little meaning.
The introduction *has* a great deal of thought *content*.
Your words seem *hollow*.
The sentence is *without* meaning.
The idea is *buried in* terribly dense paragraphs.

In examples like these it is far more difficult to see that there is anything hidden by the metaphor or even to see that there is a metaphor here at all. This is so much the conventional way of thinking about language that it is sometimes hard to imagine that it might not fit reality. But if we look at what the CONDUIT metaphor entails, we can see some of the ways in which it masks aspects of the communicative process.

First, the LINGUISTIC EXPRESSIONS ARE CONTAINERS FOR MEANINGS aspect of the CONDUIT metaphor entails that words and sentences have meanings in themselves, independent of any context or speaker. The MEANINGS ARE OBJECTS part of the metaphor, for example, entails that meanings have an existence independent of people and contexts. The part of the metaphor that says LINGUISTIC EXPRESSIONS ARE CONTAINERS FOR MEANING entails that words (and sentences) have meanings, again independent of contexts and speakers. These metaphors are appropriate in many situations—those where context differences don't matter and where all the participants in the conversation understand the sentences in the same way. These two entailments are exemplified by sentences like

The meaning is *right there in* the words,

which, according to the CONDUIT metaphor, can correctly be said of any sentence. But there are many cases where context does matter. Here is a celebrated one recorded in actual conversation by Pamela Downing:

Please sit in the apple-juice seat.

In isolation this sentence has no meaning at all, since the expression "apple-juice seat" is not a conventional way of referring to any kind of object. But the sentence makes perfect sense in the context in which it was uttered. An overnight guest came down to breakfast. There were four place settings, three with orange juice and one with apple juice. It was clear what the apple-juice seat was. And even the next morning, when there was no apple juice, it was still clear which seat was the apple-juice seat.

In addition to sentences that have no meaning without context, there are cases where a single sentence will mean different things to different people. Consider:

We need new alternative sources of energy.

This means something very different to the president of Mobil Oil from what it means to the president of Friends of the Earth. The meaning is not right there in the sentence—it matters a lot who is saying or listening to the sentence and what his social and political attitudes are. The CONDUIT metaphor does not fit cases where context is required to determine whether the sentence has any meaning at all and, if so, what meaning it has.

These examples show that the metaphorical concepts we have looked at provide us with a partial understanding of what communication, argument, and time are and that, in doing this, they hide other aspects of these concepts. It is important to see that the metaphorical structuring involved here is partial, not total. If it were total, one concept would actually *be* the other, not merely be understood in terms of it. For example, time isn't really money. If you *spend your time* trying to do something and it doesn't work, you can't get your time back. There are no time banks. I can *give you a lot of time,* but you can't give me back the same time, though you can *give me back the same amount of time.* And so on. Thus, part of a metaphorical concept does not and cannot fit.

On the other hand, metaphorical concepts can be extended beyond the range of ordinary literal ways of thinking and talking into the range of what is called figurative, poetic, colorful, or fanciful thought and language. Thus, if ideas are objects, we can *dress them up in fancy clothes, juggle them, line them up nice and neat,* etc. So when we say that a concept is structured by a metaphor, we mean that it is partially structured and that it can be extended in some ways but not others.

4

Orientational Metaphors

So far we have examined what we will call *structural metaphors,* cases where one concept is metaphorically structured in terms of another. But there is another kind of metaphorical concept, one that does not structure one concept in terms of another but instead organizes a whole system of concepts with respect to one another. We will call these *orientational metaphors,* since most of them have to do with spatial orientation: up-down, in-out, front-back, on-off, deep-shallow, central-peripheral. These spatial orientations arise from the fact that we have bodies of the sort we have and that they function as they do in our physical environment. Orientational metaphors give a concept a spatial orientation; for example, HAPPY IS UP. The fact that the concept HAPPY is oriented UP leads to English expressions like "I'm feeling *up* today."

Such metaphorical orientations are not arbitrary. They have a basis in our physical and cultural experience. Though the polar oppositions up-down, in-out, etc., are physical in nature, the orientational metaphors based on them can vary from culture to culture. For example, in some cultures the future is in front of us, whereas in others it is in back. We will be looking at up-down spatialization metaphors, which have been studied intensively by William Nagy (1974), as an illustration. In each case, we will give a brief hint about how each metaphorical concept might have arisen from our physical and cultural experience. These accounts are meant to be suggestive and plausible, not definitive.

HAPPY IS UP; SAD IS DOWN
I'm feeling *up*. That *boosted* my spirits. My spirits *rose*.
You're in *high* spirits. Thinking about her always gives me a
lift. I'm feeling *down*. I'm *depressed*. He's really *low* these
days. I *fell* into a depression. My spirits *sank*.

Physical basis: Drooping posture typically goes along
with sadness and depression, erect posture with a positive
emotional state.

CONSCIOUS IS UP; UNCONSCIOUS IS DOWN
Get *up*. Wake *up*. I'm *up* already. He *rises* early in the
morning. He *fell* asleep. He *dropped* off to sleep. He's *under*
hypnosis. He *sank* into a coma.

Physical basis: Humans and most other mammals sleep
lying down and stand up when they awaken.

HEALTH AND LIFE ARE UP; SICKNESS AND DEATH ARE DOWN
He's at the *peak* of health. Lazarus *rose* from the dead. He's
in *top* shape. As to his health, he's way *up* there. He *fell* ill.
He's *sinking* fast. He came *down* with the flu. His health is
declining. He *dropped* dead.

Physical basis: Serious illness forces us to lie down
physically. When you're dead, you are physically down.

HAVING CONTROL or FORCE IS UP; BEING SUBJECT TO CONTROL
or FORCE IS DOWN
I have control *over* her. I am *on top of* the situation. He's in a
superior position. He's at the *height* of his power. He's in the
high command. He's in the *upper* echelon. His power *rose*.
He ranks *above* me in strength. He is *under* my control. He
fell from power. His power is on the *decline*. He is my social
inferior. He is *low man* on the totem pole.

Physical basis: Physical size typically correlates with
physical strength, and the victor in a fight is typically on
top.

MORE IS UP; LESS IS DOWN
The number of books printed each year keeps going *up*. His

draft number is *high*. My income *rose* last year. The amount
of artistic activity in this state has gone *down* in the past year.
The number of errors he made is incredibly *low*. His income
fell last year. He is *under*age. If you're too hot, turn the heat
down.

Physical basis: If you add more of a substance or of
physical objects to a container or pile, the level goes up.

FORESEEABLE FUTURE EVENTS ARE UP (and AHEAD)
    All *up*coming events are listed in the paper. What's coming
    up this week? I'm afraid of what's *up ahead* of us. What's
    up?

Physical basis: Normally our eyes look in the direction in
which we typically move (ahead, forward). As an object
approaches a person (or the person approaches the object),
the object appears larger. Since the ground is perceived as
being fixed, the top of the object appears to be moving
upward in the person's field of vision.

HIGH STATUS IS UP; LOW STATUS IS DOWN
    He has a *lofty* position. She'll *rise* to the *top*. He's at the *peak*
    of his career. He's *climbing* the ladder. He has little *upward*
    mobility. He's at the *bottom* of the social hierarchy. She *fell*
    in status.

Social and physical basis: Status is correlated with (social) power and (physical) power is UP.

GOOD IS UP; BAD IS DOWN
    Things are looking *up*. We hit a *peak* last year, but it's been
    downhill ever since. Things are at an all-time *low*. He does
    high-quality work.

Physical basis for personal well-being: Happiness,
health, life, and control—the things that principally
characterize what is good for a person—are all UP.

VIRTUE IS UP; DEPRAVITY IS DOWN
    He is *high*-minded. She has *high* standards. She is *upright*.
    She is an *upstanding* citizen. That was a *low* trick. Don't be

underhanded. I wouldn't *stoop* to that. That would be *beneath* me. He *fell* into the *abyss* of depravity. That was a *low-down* thing to do.

Physical and social basis: GOOD IS UP for a person (physical basis), together with a metaphor that we will discuss below, SOCIETY IS A PERSON (in the version where you are *not* identifying with your society). To be virtuous is to act in accordance with the standards set by the society/person to maintain its well-being. VIRTUE IS UP because virtuous actions correlate with social well-being from the society/person's point of view. Since socially based metaphors are part of the culture, it's the society/person's point of view that counts.

RATIONAL IS UP; EMOTIONAL IS DOWN

The discussion *fell to the emotional* level, but I *raised* it back *up to the rational* plane. We put our *feelings* aside and had a *high-level intellectual* discussion of the matter. He couldn't *rise above* his *emotions.*

Physical and cultural basis: In our culture people view themselves as being in control over animals, plants, and their physical environment, and it is their unique ability to reason that places human beings above other animals and gives them this control. CONTROL IS UP thus provides a basis for MAN IS UP and therefore for RATIONAL IS UP.

Conclusions

On the basis of these examples, we suggest the following conclusions about the experiential grounding, the coherence, and the systematicity of metaphorical concepts:

—Most of our fundamental concepts are organized in terms of one or more spatialization metaphors.

—There is an internal systematicity to each spatialization metaphor. For example, HAPPY IS UP defines a coherent system rather than a number of isolated and random cases. (An example of an incoherent system would be one where, say, "I'm feeling up" meant "I'm feeling happy," but "My spirits rose" meant "I became sadder.")

—There is an overall external systematicity among the various spatialization metaphors, which defines coherence among them. Thus, GOOD IS UP gives an UP orientation to general well-being, and this orientation is coherent with special cases like HAPPY IS UP, HEALTH IS UP, ALIVE IS UP, CONTROL IS UP. STATUS IS UP is coherent with CONTROL IS UP.

—Spatialization metaphors are rooted in physical and cultural experience; they are not randomly assigned. A metaphor can serve as a vehicle for understanding a concept only by virtue of its experiential basis. (Some of the complexities of the experiential basis of metaphor are discussed in the following section.)

—There are many possible physical and social bases for metaphor. Coherence within the overall system seems to be part of the reason why one is chosen and not another. For example, happiness also tends to correlate physically with a smile and a general feeling of expansiveness. This could in principle form the basis for a metaphor HAPPY IS WIDE; SAD IS NARROW. And in fact there are minor metaphorical expressions, like "I'm feeling *expansive*," that pick out a different aspect of happiness than "I'm feeling *up*" does. But the major metaphor in our culture is HAPPY IS UP; there is a reason why we speak of the height of ecstasy rather than the breadth of ecstasy. HAPPY IS UP is maximally coherent with GOOD IS UP, HEALTHY IS UP, etc.

—In some cases spatialization is so essential a part of a concept that it is difficult for us to imagine any alternative metaphor that might structure the concept. In our society "high status" is such a concept. Other cases, like happiness, are less clear. Is the concept of happiness independent of the HAPPY IS UP metaphor, or is the up-down spatialization of happiness a part of the concept? We believe that it is a part of the concept within a given conceptual system. The HAPPY IS UP metaphor places happiness within a coherent metaphorical system, and part of its meaning comes from its role in that system.

—So-called purely intellectual concepts, e.g., the concepts in a scientific theory, are often—perhaps always—based on metaphors that have a physical and/or cultural basis. The *high* in "high-energy particles" is based on MORE IS UP. The *high* in "high-level functions," as in physiological psychology, is based on RATIONAL IS UP. The *low* in "low-level phonology" (which refers to detailed phonetic aspects of the sound systems of languages) is based on MUNDANE REALITY IS DOWN (as in "down to earth"). The intuitive appeal of a scientific theory has to do with how well its metaphors fit one's experience.

—Our physical and cultural experience provides many possible bases for spatialization metaphors. Which ones are chosen, and which ones are major, may vary from culture to culture.

—It is hard to distinguish the physical from the cultural basis of a metaphor, since the choice of one physical basis from among many possible ones has to do with cultural coherence.

Experiential Bases of Metaphors

We do not know very much about the experiential bases of metaphors. Because of our ignorance in this matter, we have described the metaphors separately, only later adding speculative notes on their possible experiential bases. We are adopting this practice out of ignorance, not out of principle. *In actuality we feel that no metaphor can ever be comprehended or even adequately represented independently of its experiential basis.* For example, MORE IS UP has a very different kind of experiential basis than HAPPY IS UP or RATIONAL IS UP. Though the concept UP is the same in all these metaphors, the experiences on which these UP metaphors are based are very different. It is not that there are many different UPs; rather, verticality enters our experience in many different ways and so gives rise to many different metaphors.

One way of emphasizing the inseparability of metaphors from their experiential bases would be to build the experiential basis into the representations themselves. Thus, instead of writing MORE IS UP and RATIONAL IS UP, we might have the more complex relationship shown in the diagram.

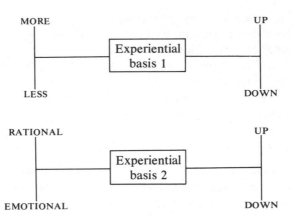

Such a representation would emphasize that the two parts of each metaphor are linked only via an experiential basis and that it is only by means of these experiential bases that the metaphor can serve the purpose of understanding.

We will not use such representations, but only because we know so little about experiential bases of metaphors. We will continue to use the word "is" in stating metaphors like MORE IS UP, but the IS should be viewed as a shorthand for some set of experiences on which the metaphor is based and in terms of which we understand it.

The role of the experiential basis is important in understanding the workings of metaphors that do not fit together because they are based on different kinds of experience. Take, for example, a metaphor like UNKNOWN IS UP; KNOWN IS DOWN. Examples are "That's *up in the air*" and "The matter is *settled.*" This metaphor has an experiential basis very much like that of UNDERSTANDING IS GRASPING, as in "I couldn't *grasp* his explanation." With physical objects, if you can grasp something and hold it in your hands, you can look it over carefully and get a reasonably good understanding of it. It's easier to grasp something and look at it carefully if it's on the ground in a fixed location than if it's floating through the air (like a leaf or a piece of paper). Thus UNKNOWN IS UP; KNOWN IS DOWN is coherent with UNDERSTANDING IS GRASPING.

But UNKNOWN IS UP is not coherent with metaphors like GOOD IS UP and FINISHED IS UP (as in "I'm finishing *up*"). One would expect FINISHED to be paired with KNOWN and UNFINISHED to be paired with UNKNOWN. But, so far as verticality metaphors are concerned, this is not the case. The reason is that UNKNOWN IS UP has a very different experiential basis than FINISHED IS UP.

5

Metaphor and Cultural Coherence

The most fundamental values in a culture will be coherent with the metaphorical structure of the most fundamental concepts in the culture. As an example, let us consider some cultural values in our society that are coherent with our UP-DOWN spatialization metaphors and whose opposites would not be.

"More is better" is coherent with MORE IS UP and GOOD IS UP. "Less is better" is not coherent with them.

"Bigger is better" is coherent with MORE IS UP and GOOD IS UP. "Smaller is better" is not coherent with them.

"The future will be better" is coherent with THE FUTURE IS UP and GOOD IS UP. "The future will be worse" is not.

"There will be more in the future" is coherent with MORE IS UP and THE FUTURE IS UP.

"Your status should be higher in the future" is coherent with HIGH STATUS IS UP and THE FUTURE IS UP.

These are values deeply embedded in our culture. "The future will be better" is a statement of the concept of progress. "There will be more in the future" has as special cases the accumulation of goods and wage inflation. "Your status should be higher in the future" is a statement of careerism. These are coherent with our present spatialization metaphors; their opposites would not be. So it seems that our values are not independent but must form a coherent system with the metaphorical concepts we live by. We are not claiming that all cultural values coherent with a metaphorical system actually exist, only that those that do exist and are deeply entrenched are consistent with the metaphorical system.

The values listed above hold in our culture generally—all things being equal. But because things are usually not equal, there are often conflicts among these values and hence conflicts among the metaphors associated with them. To explain such conflicts among values (and their metaphors), we must find the different priorities given to these values and metaphors by the subculture that uses them. For instance, MORE IS UP seems always to have the highest priority since it has the clearest physical basis. The priority of MORE IS UP over GOOD IS UP can be seen in examples like "Inflation is rising" and "The crime rate is going up." Assuming that inflation and the crime rate are bad, these sentences mean what they do because MORE IS UP always has top priority.

In general, which values are given priority is partly a matter of the subculture one lives in and partly a matter of personal values. The various subcultures of a mainstream culture share basic values but give them different priorities. For example, BIGGER IS BETTER may be in conflict with THERE WILL BE MORE IN THE FUTURE when it comes to the question of whether to buy a big car now, with large time payments that will eat up future salary, or whether to buy a smaller, cheaper car. There are American subcultures where you buy the big car and don't worry about the future, and there are others where the future comes first and you buy the small car. There was a time (before inflation and the energy crisis) when owning a small car had a high status within the subculture where VIRTUE IS UP and SAVING RESOURCES IS VIRTUOUS took priority over BIGGER IS BETTER. Nowadays the number of small-car owners has gone up drastically because there is a large subculture where SAVING MONEY IS BETTER has priority over BIGGER IS BETTER.

In addition to subcultures, there are groups whose defining characteristic is that they share certain important values that conflict with those of the mainstream culture. But in less obvious ways they preserve other mainstream values. Take monastic orders like the Trappists. There LESS IS BETTER and SMALLER IS BETTER are true with respect to material possessions, which are viewed as hindering what is important, namely, serving God. The Trappists share the mainstream value VIRTUE IS UP, though they give it the highest priority and a very different definition. MORE is still BETTER, though it applies to virtue; and status is still UP, though it is not of this world but of a higher one, the Kingdom of God. Moreover, THE FUTURE WILL BE BETTER is true in terms of spiritual growth (UP) and, ultimately, salvation (really UP). This is typical of groups that are out of the mainstream culture. Virtue, goodness, and status may be radically redefined, but they are still UP. It is still better to have more of what is important, THE FUTURE WILL BE BETTER with respect to what is important, and so on. Relative to what is important for a monastic group, the value system is both internally coherent and, with respect to what is important for the group, coherent with the major orientational metaphors of the mainstream culture.

Individuals, like groups, vary in their priorities and in the ways they define what is good or virtuous to them. In this sense, they are subgroups of one. Relative to what is important for them, their individual value systems are coherent with the major orientational metaphors of the mainstream culture.

Not all cultures give the priorities we do to up-down orientation. There are cultures where balance or centrality plays a much more important role than it does in our culture. Or consider the nonspatial orientation active-passive. For us ACTIVE IS UP and PASSIVE IS DOWN in most matters. But there are cultures where passivity is valued more than activity. In general the major orientations up-down, in-out, central-peripheral, active-passive, etc., seem to cut across all cultures, but which concepts are oriented which way and which orientations are most important vary from culture to culture.

6

Ontological Metaphors

Entity and Substance Metaphors

Spatial orientations like up-down, front-back, on-off, center-periphery, and near-far provide an extraordinarily rich basis for understanding concepts in orientational terms. But one can do only so much with orientation. Our experience of physical objects and substances provides a further basis for understanding—one that goes beyond mere orientation. Understanding our experiences in terms of objects and substances allows us to pick out parts of our experience and treat them as discrete entities or substances of a uniform kind. Once we can identify our experiences as entities or substances, we can refer to them, categorize them, group them, and quantify them—and, by this means, reason about them.

When things are not clearly discrete or bounded, we still categorize them as such, e.g., mountains, street corners, hedges, etc. Such ways of viewing physical phenomena are needed to satisfy certain purposes that we have: locating mountains, meeting at street corners, trimming hedges. Human purposes typically require us to impose artificial boundaries that make physical phenomena discrete just as we are: entities bounded by a surface.

Just as the basic experiences of human spatial orientations give rise to orientational metaphors, so our experiences with physical objects (especially our own bodies) provide the basis for an extraordinarily wide variety of ontological metaphors, that is, ways of viewing events, activities, emotions, ideas, etc., as entities and substances.

Ontological metaphors serve various purposes, and the

25

various kinds of metaphors there are reflect the kinds of purposes served. Take the experience of rising prices, which can be metaphorically viewed as an entity via the noun *inflation*. This gives us a way of referring to the experience:

INFLATION IS AN ENTITY

Inflation is lowering our standard of living.
If there's much *more inflation,* we'll never survive.
We need to *combat inflation.*
Inflation is backing us into a corner.
Inflation is taking its toll at the checkout counter and the gas pump.
Buying land is the best way of *dealing with inflation.*
Inflation makes me sick.

In these cases, viewing inflation as an entity allows us to refer to it, quantify it, identify a particular aspect of it, see it as a cause, act with respect to it, and perhaps even believe that we understand it. Ontological metaphors like this are necessary for even attempting to deal rationally with our experiences.

The range of ontological metaphors that we use for such purposes is enormous. The following list gives some idea of the kinds of purposes, along with representative examples of ontological metaphors that serve them.

Referring
My *fear of insects* is driving my wife crazy.
That was a *beautiful catch.*
We are working toward *peace.*
The *middle class* is a *powerful silent force* in *American politics.*
The *honor of our country* is at stake in this war.

Quantifying
It will take *a lot of patience* to finish this book.
There is *so much hatred* in the world.
DuPont has *a lot of political power* in Delaware.
You've got *too much hostility* in you.

Pete Rose has *a lot of hustle and baseball know-how*.

Identifying Aspects
The *ugly side of his personality* comes out under pressure.
The *brutality of war* dehumanizes us all.
I can't keep up with the *pace of modern life*.
His *emotional health* has deteriorated recently.
We never got to feel the *thrill of victory* in Vietnam.

Identifying Causes
The *pressure of his responsibilities* caused his breakdown.
He did it out of *anger*.
Our influence in the world has declined because of our *lack of moral fiber*.
Internal dissension cost them the pennant.

Setting Goals and Motivating Actions
He went to New York to *seek fame and fortune*.
Here's what you have to do to *insure financial security*.
I'm changing my way of life so that I can *find true happiness*.
The FBI will act quickly in the face of a *threat to national security*.
She saw getting married as the *solution to her problems*.

As in the case of orientational metaphors, most of these expressions are not noticed as being metaphorical. One reason for this is that ontological metaphors, like orientational metaphors, serve a very limited range of purposes—referring, quantifying, etc. Merely viewing a nonphysical thing as an entity or substance does not allow us to comprehend very much about it. But ontological metaphors may be further elaborated. Here are two examples of how the ontological metaphor THE MIND IS AN ENTITY is elaborated in our culture.

THE MIND IS A MACHINE
We're still trying to *grind out* the solution to this equation.
My mind just isn't *operating* today.
Boy, the *wheels are turning* now!
I'm *a little rusty* today.
We've been working on this problem all day and now we're *running out of steam*.

THE MIND IS A BRITTLE OBJECT
Her ego is very *fragile*.
You have to *handle him with care* since his wife's death.
He *broke* under cross-examination.
She is *easily crushed*.
The experience *shattered* him.
I'm *going to pieces*.
His mind *snapped*.

These metaphors specify different kinds of objects. They give us different metaphorical models for what the mind is and thereby allow us to focus on different aspects of mental experience. The MACHINE metaphor gives us a conception of the mind as having an on-off state, a level of efficiency, a productive capacity, an internal mechanism, a source of energy, and an operating condition. The BRITTLE OBJECT metaphor is not nearly as rich. It allows us to talk only about psychological strength. However, there is a range of mental experience that can be conceived of in terms of either metaphor. The examples we have in mind are these:

He broke down. (THE MIND IS A MACHINE)
He cracked up. (THE MIND IS A BRITTLE OBJECT)

But these two metaphors do not focus on *exactly* the same aspect of mental experience. When a machine breaks down, it simply ceases to function. When a brittle object shatters, its pieces go flying, with possibly dangerous consequences. Thus, for example, when someone goes crazy and becomes wild or violent, it would be appropriate to say "He cracked up." On the other hand, if someone becomes lethargic and unable to function for psychological reasons, we would be more likely to say "He broke down."

Ontological metaphors like these are so natural and so pervasive in our thought that they are usually taken as self-evident, direct descriptions of mental phenomena. The fact that they are metaphorical never occurs to most of us. We take statements like "He cracked under pressure" as being directly true or false. This expression was in fact used by various journalists to explain why Dan White brought his gun to the San Francisco City Hall and shot and killed Mayor George Moscone. Explanations of this sort seem perfectly natural to most of us. The reason is that metaphors like THE MIND IS A BRITTLE OBJECT are an integral part of the model of the mind that we have in this culture; it is the model most of us think and operate in terms of.

Container Metaphors

Land Areas

We are physical beings, bounded and set off from the rest of the world by the surface of our skins, and we experience the rest of the world as outside us. Each of us is a container, with a bounding surface and an in-out orientation. We project our own in-out orientation onto other physical objects that are bounded by surfaces. Thus we also view them as containers with an inside and an outside. Rooms and houses are obvious containers. Moving from room to room is moving from one container to another, that is, moving *out of* one room and *into* another. We even give solid objects this orientation, as when we break a rock open to see what's inside it. We impose this orientation on our natural environment as well. A clearing in the woods is seen as having a bounding surface, and we can view ourselves as being *in* the clearing or *out of* the clearing, *in* the woods or *out of* the woods. A clearing in the woods has something we can perceive as a natural boundary—the fuzzy area where the trees more or less stop and the clearing more or less begins. But even where there is no natural physical boundary that can be viewed as defining a container, we impose boundaries—marking off territory so that it has an inside and a bounding surface—whether a wall, a fence, or an abstract line or plane. There are few human instincts more basic than territoriality. And such defining of a territory, putting a boundary around it, is an act of quantification.

Bounded objects, whether human beings, rocks, or land
areas, have sizes. This allows them to be quantified in terms
of the amount of substance they contain. Kansas, for
example, is a bounded area— a CONTAINER—which is why
we can say, "There's a lot of land *in* Kansas."

Substances can themselves be viewed as containers.
Take a tub of water, for example. When you get into the
tub, you get into the water. Both the tub and the water are
viewed as containers, but of different sorts. The tub is a
CONTAINER OBJECT, while the water is a CONTAINER SUB-
STANCE.

The Visual Field

We conceptualize our visual field as a container and conceptualize what we see as being inside it. Even the term
"visual *field*" suggests this. The metaphor is a natural one
that emerges from the fact that, when you look at some
territory (land, floor space, etc.), your field of vision defines
a boundary of the territory, namely, the part that you can
see. Given that a bounded physical space is a CONTAINER
and that our field of vision correlates with that bounded
physical space, the metaphorical concept VISUAL FIELDS
ARE CONTAINERS emerges naturally. Thus we can say:

> The ship is *coming into* view.
> I *have* him *in* sight.
> I can't see him—the tree is *in* the way.
> He's *out of* sight now.
> That's *in* the *center of* my *field* of vision.
> There's *nothing in* sight.
> I can't get *all* of the ships *in* sight at once.

Events, Actions, Activities, and States

We use ontological metaphors to comprehend events, actions, activities, and states. Events and actions are conceptualized metaphorically as objects, activities as substances, states as containers. A race, for example, is an
event, which is viewed as a discrete entity. The race exists

in space and time, and it has well-defined boundaries. Hence we view it as a CONTAINER OBJECT, having in it participants (which are objects), events like the start and finish (which are metaphorical objects), and the activity of running (which is a metaphorical substance). Thus we can say of a race:

> Are you *in* the race on Sunday? (race as CONTAINER OBJECT)
> Are you *going to* the race? (race as OBJECT)
> Did you *see* the race? (race as OBJECT)
> The *finish* of the race was really exciting. (finish as EVENT OBJECT within CONTAINER OBJECT)
> There was *a lot of good running in* the race. (running as a SUBSTANCE in a CONTAINER)
> I couldn't do *much sprinting* until the end. (sprinting as SUBSTANCE)
> *Halfway into* the race, I ran out of energy. (race as CONTAINER OBJECT)
> He's *out of* the race now. (race as CONTAINER OBJECT)

Activities in general are viewed metaphorically as SUBSTANCES and therefore as CONTAINERS:

> *In* washing the window, I splashed water all over the floor.
> How did Jerry *get out of* washing the windows?
> *Outside of* washing the windows, what else did you do?
> *How much* window-washing did you do?
> How did you *get into* window-washing as a profession?
> He's *immersed in* washing the windows right now.

Thus, activities are viewed as containers for the actions and other activities that make them up. They are also viewed as containers for the energy and materials required for them and for their by-products, which may be viewed as *in* them or as *emerging from* them:

> I *put a lot of energy into* washing the windows.
> I *get a lot of satisfaction out of* washing windows.
> *There is a lot of satisfaction in* washing windows.

Various kinds of states may also be conceptualized as containers. Thus we have examples like these:

He's *in* love.
We're *out of* trouble now.
He's *coming out of* the coma.
I'm *slowly getting into* shape.
He *entered* a state of euphoria.
He *fell into* a depression.
He finally *emerged from* the catatonic state he had been *in* since the end of finals week.

7

Personification

Perhaps the most obvious ontological metaphors are those where the physical object is further specified as being a person. This allows us to comprehend a wide variety of experiences with nonhuman entities in terms of human motivations, characteristics, and activities. Here are some examples:

His *theory explained* to me the behavior of chickens raised in factories.
This *fact argues* against the standard theories.
Life has cheated me.
Inflation is eating up our profits.
His *religion tells* him that he cannot drink fine French wines.
The *Michelson-Morley experiment gave birth to* a new physical theory.
Cancer finally *caught up with* him.

In each of these cases we are seeing something nonhuman as human. But personification is not a single unified general process. Each personification differs in terms of the aspects of people that are picked out. Consider these examples.

Inflation *has attacked* the foundation of our economy.
Inflation *has pinned* us *to the wall*.
Our biggest *enemy* right now *is* inflation.
The dollar *has been destroyed* by inflation.
Inflation *has robbed* me of my savings.
Inflation *has outwitted* the best economic minds in the country.
Inflation *has given birth* to a money-minded generation.

Here inflation is personified, but the metaphor is not

merely INFLATION IS A PERSON. It is much more specific, namely, INFLATION IS AN ADVERSARY. It not only gives us a very specific way of thinking about inflation but also a way of acting toward it. We think of inflation as an adversary that can attack us, hurt us, steal from us, even destroy us. The INFLATION IS AN ADVERSARY metaphor therefore gives rise to and justifies political and economic actions on the part of our government: declaring war on inflation, setting targets, calling for sacrifices, installing a new chain of command, etc.

The point here is that personification is a general category that covers a very wide range of metaphors, each picking out different aspects of a person or ways of looking at a person. What they all have in common is that they are extensions of ontological metaphors and that they allow us to make sense of phenomena in the world in human terms—terms that we can understand on the basis of our own motivations, goals, actions, and characteristics. Viewing something as abstract as inflation in human terms has an explanatory power of the only sort that makes sense to most people. When we are suffering substantial economic losses due to complex economic and political factors that no one really understands, the INFLATION IS AN ADVERSARY metaphor at least gives us a coherent account of why we're suffering these losses.

8

Metonymy

In the cases of personification that we have looked at we are imputing human qualities to things that are not human— theories, diseases, inflation, etc. In such cases there are no actual human beings referred to. When we say "Inflation robbed me of my savings," we are not using the term "inflation" to refer to a person. Cases like this must be distinguished from cases like

The *ham sandwich* is waiting for his check.

where the expression "the ham sandwich" is being used to refer to an actual person, the person who ordered the ham sandwich. Such cases are not instances of personification metaphors, since we do not understand "the ham sandwich" by imputing human qualities to it. Instead, we are using one entity to refer to another that is related to it. This is a case of what we will call *metonymy*. Here are some further examples:

He likes to read the *Marquis de Sade*. (= the writings of the marquis)
He's in *dance*. (= the dancing profession)
Acrylic has taken over the art world. (= the use of acrylic paint)
The *Times* hasn't arrived at the press conference yet. (= the reporter from the *Times*)
Mrs. Grundy frowns on *blue jeans*. (= the wearing of blue jeans)
New windshield wipers will satisfy him. (= the state of having new wipers)

We are including as a special case of metonymy what traditional rhetoricians have called *synecdoche,* where the part
stands for the whole, as in the following.

THE PART FOR THE WHOLE

> The *automobile* is clogging our highways. (= the collection
> of automobiles)
> We need a couple of *strong bodies* for our team. (= strong
> people)
> There are a lot of *good heads* in the university. (= intelligent
> people)
> I've got a new *set of wheels.* (= car, motorcycle, etc.)
> We need some *new blood* in the organization. (= new people)

In these cases, as in the other cases of metonymy, one
entity is being used to refer to another. Metaphor and
metonymy are different *kinds* of processes. Metaphor is
principally a way of conceiving of one thing in terms of
another, and its primary function is understanding.
Metonymy, on the other hand, has primarily a referential
function, that is, it allows us to use one entity to *stand for*
another. But metonymy is not merely a referential device.
It also serves the function of providing understanding. For
example, in the case of the metonymy THE PART FOR THE
WHOLE there are many parts that can stand for the whole.
Which part we pick out determines which aspect of the
whole we are focusing on. When we say that we need some
good heads on the project, we are using "good heads" to
refer to "intelligent people." The point is not just to use a
part (head) to stand for a whole (person) but rather to pick
out a particular characteristic of the person, namely, intelligence, which is associated with the head. The same is true
of other kinds of metonymies. When we say "The *Times*
hasn't arrived at the press conference yet," we are using
"The *Times*" not merely to refer to some reporter or other
but also to suggest the importance of the institution the
reporter represents. So "The *Times* has not yet arrived for
the press conference" means something different from

"Steve Roberts has not yet arrived for the press conference," even though Steve Roberts may be the *Times* reporter in question.

Thus metonymy serves some of the same purposes that metaphor does, and in somewhat the same way, but it allows us to focus more specifically on certain aspects of what is being referred to. It is also like metaphor in that it is not just a poetic or rhetorical device. Nor is it just a matter of language. Metonymic concepts (like THE PART FOR THE WHOLE) are part of the ordinary, everyday way we think and act as well as talk.

For example, we have in our conceptual system a special case of the metonymy THE PART FOR THE WHOLE, namely, THE FACE FOR THE PERSON. For example:

> She's just a *pretty face*.
> There are an *awful lot of faces* out there in the audience.
> We need some *new faces* around here.

This metonymy functions actively in our culture. The tradition of portraits, in both painting and photography, is based on it. If you ask me to show you a picture of my son and I show you a picture of his face, you will be satisfied. You will consider yourself to have seen a picture of him. But if I show you a picture of his body without his face, you will consider it strange and will not be satisfied. You might even ask, "But what does he look like?" Thus the metonymy THE FACE FOR THE PERSON is not merely a matter of language. In our culture we look at a person's face—rather than his posture or his movements—to get our basic information about what the person is like. We function in terms of a metonymy when we perceive the person in terms of his face and act on those perceptions.

Like metaphors, metonymies are not random or arbitrary occurrences, to be treated as isolated instances. Metonymic concepts are also systematic, as can be seen in the following representative examples that exist in our culture.

THE PART FOR THE WHOLE

Get *your butt* over here!
We don't hire *longhairs*.
The Giants need a *stronger arm* in right field.
I've got a new *four-on-the-floor V-8*.

PRODUCER FOR PRODUCT

I'll have a *Löwenbräu*.
He bought a *Ford*.
He's got a *Picasso* in his den.
I hate to read *Heidegger*.

OBJECT USED FOR USER

The *sax* has the flu today.
The *BLT* is a lousy tipper.
The *gun* he hired wanted fifty grand.
We need a better *glove* at third base.
The *buses* are on strike.

CONTROLLER FOR CONTROLLED

Nixon bombed Hanoi.
Ozawa gave a terrible concert last night.
Napoleon lost at Waterloo.
Casey Stengel won a lot of pennants.
A Mercedes rear-ended *me*.

INSTITUTION FOR PEOPLE RESPONSIBLE

Exxon has raised its prices again.
You'll never get the *university* to agree to that.
The *Army* wants to reinstitute the draft.
The *Senate* thinks abortion is immoral.
I don't approve of the *government's* actions.

THE PLACE FOR THE INSTITUTION

The *White House* isn't saying anything.
Washington is insensitive to the needs of the people.
The *Kremlin* threatened to boycott the next round of SALT
  talks.
Paris is introducing longer skirts this season.
Hollywood isn't what it used to be.
Wall Street is in a panic.

THE PLACE FOR THE EVENT
Let's not let Thailand become another *Vietnam*.
Remember *the Alamo*.
Pearl Harbor still has an effect on our foreign policy.
Watergate changed our politics.
It's been *Grand Central Station* here all day.

Metonymic concepts like these are systematic in the same way that metaphoric concepts are. The sentences given above are not random. They are instances of certain general metonymic concepts in terms of which we organize our thoughts and actions. Metonymic concepts allow us to conceptualize one thing by means of its relation to something else. When we think of *a Picasso,* we are not just thinking of a work of art alone, in and of itself. We think of it in terms of its relation to the artist, that is, his conception of art, his technique, his role in art history, etc. We act with reverence toward *a Picasso,* even a sketch he made as a teen-ager, because of its relation to the artist. This is a way in which the PRODUCER FOR PRODUCT metonymy affects both our thought and our action. Similarly, when a waitress says "The ham sandwich wants his check," she is not interested in the person as a person but only as a customer, which is why the use of such a sentence is dehumanizing. Nixon himself may not have dropped the bombs on Hanoi, but via the CONTROLLER FOR CONTROLLED metonymy we not only say "Nixon bombed Hanoi" but also think of him as doing the bombing and hold him responsible for it. Again this is possible because of the nature of the metonymic relationship in the CONTROLLER FOR CONTROLLED metonymy, where responsibility is what is focused on.

Thus, like metaphors, metonymic concepts structure not just our language but our thoughts, attitudes, and actions. And, like metaphoric concepts, metonymic concepts are grounded in our experience. In fact, the grounding of metonymic concepts is in general more obvious than is the case with metaphoric concepts, since it usually involves direct physical or causal associations. The PART FOR

WHOLE metonymy, for example, emerges from our experiences with the way parts in general are related to wholes. PRODUCER FOR PRODUCT is based on the causal (and typically physical) relationship between a producer and his product. THE PLACE FOR THE EVENT is grounded in our experience with the physical location of events. And so on.

Cultural and religious symbolism are special cases of metonymy. Within Christianity, for example, there is the metonymy DOVE FOR HOLY SPIRIT. As is typical with metonymies, this symbolism is not arbitrary. It is grounded in the conception of the dove in Western culture and the conception of the Holy Spirit in Christian theology. There is a reason why the dove is the symbol of the Holy Spirit and not, say, the chicken, the vulture, or the ostrich. The dove is conceived of as beautiful, friendly, gentle, and, above all, peaceful. As a bird, its natural habitat is the sky, which metonymically stands for heaven, the natural habitat of the Holy Spirit. The dove is a bird that flies gracefully, glides silently, and is typically seen coming out of the sky and landing among people.

The conceptual systems of cultures and religions are metaphorical in nature. Symbolic metonymies are critical links between everyday experience and the coherent metaphorical systems that characterize religions and cultures. Symbolic metonymies that are grounded in our physical experience provide an essential means of comprehending religious and cultural concepts.

9

Challenges to Metaphorical Coherence

We have offered evidence that metaphors and metonymies are not random but instead form coherent systems in terms of which we conceptualize our experience. But it is easy to find apparent incoherences in everyday metaphorical expressions. We have not made a complete study of these, but those that we have looked at in detail have turned out not to be incoherent at all, though they appeared that way at first. Let us consider two examples.

An Apparent Metaphorical Contradiction

Charles Fillmore has observed (in conversation) that English appears to have two contradictory organizations of time. In the first, the future is in front and the past is behind:

> In the weeks ahead of us... (future)
> That's all behind us now. (past)

In the second, the future is behind and the past is in front:

> In the following weeks... (future)
> In the preceding weeks... (past)

This appears to be a contradiction in the metaphorical organization of time. Moreover, the apparently contradictory metaphors can mix with no ill effect, as in

> We're looking *ahead* to the *following* weeks.

Here it appears that *ahead* organizes the future in front, while *following* organizes it behind.

To see that there is, in fact, a coherence here, we first

41

have to consider some facts about front-back organization. Some things, like people and cars, have inherent fronts and backs, but others, like trees, do not. A rock may receive a front-back organization under certain circumstances. Suppose you are looking at a medium-sized rock and there is a ball between you and the rock—say, a foot away from the rock. Then it is appropriate for you to say "The ball is in front of the rock." The rock has received a front-back orientation, as if it had a front that faced you. This is not universal. There are languages—Hausa, for instance—where the rock would receive the reverse orientation and you would say that the ball was behind the rock if it was between you and the rock.

Moving objects generally receive a front-back orientation so that the front is in the direction of motion (or in the canonical direction of motion, so that a car backing up retains its front). A spherical satellite, for example, that has no front while standing still, gets a front while in orbit by virtue of the direction in which it is moving.

Now, time in English is structured in terms of the TIME IS A MOVING OBJECT metaphor, with the future moving toward us:

> The time will come when...
> The time has long since gone when...
> The time for action has arrived.

The proverb "Time flies" is an instance of the TIME IS A MOVING OBJECT metaphor. Since we are facing toward the future, we get:

> Coming up in the weeks ahead...
> I look forward to the arrival of Christmas.
> Before us is a great opportunity, and we don't want it to pass us by.

By virtue of the TIME IS A MOVING OBJECT metaphor, time receives a front-back orientation facing in the direction of motion, just as any moving object would. Thus the future is facing toward us as it moves toward us, and we find expressions like:

I can't face the future.
The face of things to come . . .
Let's meet the future head-on.

Now, while expressions like *ahead of us, I look forward,* and *before us* orient times with respect to people, expressions like *precede* and *follow* orient times with respect to times. Thus we get:

Next week and the week following it.

but not:

The week following me . . .

Since future times are facing toward us, the times following them are further in the future, and all future times follow the present. That is why the *weeks to follow* are the same as the *weeks ahead of us.*

The point of this example is not merely to show that there is no contradiction but also to show all the subtle details that are involved: the TIME IS A MOVING OBJECT metaphor, the front-back orientation given to time by virtue of its being a moving object, and the consistent application of words like *follow, precede,* and *face* when applied to time on the basis of the metaphor. All of this consistent detailed metaphorical structure is part of our everyday literal language about time, so familiar that we would normally not notice it.

Coherence versus Consistency

We have shown that the TIME IS A MOVING OBJECT metaphor has an internal consistency. But there is another way in which we conceptualize the passing of time:

TIME IS STATIONARY AND WE MOVE THROUGH IT
As we go through the years, . . .

As we go further into the 1980s,...
We're approaching the end of the year.

What we have here are two subcases of TIME PASSES US: in one case, we are moving and time is standing still; in the other, time is moving and we are standing still. What is in common is relative motion with respect to us, with the future in front and the past behind. That is, they are two subcases of the same metaphor, as shown in the accompanying diagram.

From our point of view
time goes past us,
from front to back

Time is a moving object   Time is stationary and we
and moves toward us   move through it in the
                                           direction of the future

This is another way of saying that they have a major common entailment. Both metaphors entail that, from our point of view, time goes past us from front to back.

Although the two metaphors are not consistent (that is, they form no single image), they nonetheless "fit together," by virtue of being subcategories of a major category and therefore sharing a major common entailment. There is a difference between metaphors that are *coherent* (that is, "fit together") with each other and those that are *consistent*. We have found that the connections between metaphors are more likely to involve coherence than consistency.

As another example, let us take another metaphor:

LOVE IS A JOURNEY

Look *how far we've come.*
We're *at a crossroads.*
We'll just have to *go our separate ways.*
We can't *turn back now.*
I don't think this relationship is *going anywhere.*

Where are we?
We're *stuck*.
It's been a *long, bumpy road*.
This relationship is a *dead-end street*.
We're just *spinning our wheels*.
Our marriage is *on the rocks*.
We've gotten *off the track*.
This relationship is *foundering*.

Here the basic metaphor is that of a JOURNEY, and there are various types of journeys that one can make: a car trip, a train trip, or a sea voyage.

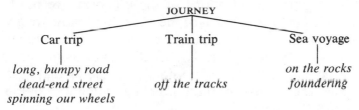

Once again, there is no single consistent image that the JOURNEY metaphors all fit. What makes them *coherent* is that they are all JOURNEY metaphors, though they specify different means of travel. The same sort of thing occurs with the TIME IS A MOVING OBJECT metaphor, where there are various ways in which something can move. Thus, *time flies, time creeps along, time speeds by.* In general, metaphorical concepts are defined not in terms of concrete images (flying, creeping, going down the road, etc.), but in terms of more general categories, like passing.

10

Some Further Examples

We have been claiming that metaphors partially structure our everyday concepts and that this structure is reflected in our literal language. Before we can get an overall picture of the philosophical implications of these claims, we need a few more examples. In each of the ones that follow we give a metaphor and a list of ordinary expressions that are special cases of the metaphor. The English expressions are of two sorts: simple literal expressions and idioms that fit the metaphor and are part of the normal everyday way of talking about the subject.

THEORIES (and ARGUMENTS) ARE BUILDINGS

Is that the *foundation* for your theory? The theory needs more *support*. The argument is *shaky*. We need some more facts or the argument will *fall apart*. We need to *construct* a *strong* argument for that. I haven't figured out yet what the *form* of the argument will be. Here are some more facts to *shore up* the theory. We need to *buttress* the theory with *solid* arguments. The theory will *stand* or *fall* on the *strength* of that argument. The argument *collapsed*. They *exploded* his latest theory. We will show that theory to be without *foundation*. So far we have put together only the *framework* of the theory.

IDEAS ARE FOOD

What he said *left a bad taste in my mouth*. All this paper has in it are *raw facts, half-baked ideas, and warmed-over theories*. There are too many facts here for me to *digest* them all. I just can't *swallow* that claim. That argument *smells fishy*. Let me *stew* over that for a while. Now there's a theory you can really *sink your teeth into*. We need to let that idea *percolate* for a while. That's *food for thought*. He's a *voracious* reader. We don't need to *spoon-feed* our students. He *devoured* the book. Let's let that idea *simmer on the back burner* for a while. This is the *meaty* part of the paper. Let that idea *jell* for a while. That idea has been *fermenting* for years.

With respect to life and death IDEAS ARE ORGANISMS, either PEOPLE or PLANTS.

IDEAS ARE PEOPLE

The theory of relativity *gave birth to* an enormous number of ideas in physics. He is the *father* of modern biology. Whose *brainchild* was that? Look at what his ideas have *spawned*. Those ideas *died off* in the Middle Ages. His ideas will *live on* forever. Cognitive psychology is still in its *infancy*. That's an idea that ought to be *resurrected*. Where'd you *dig up* that idea? He *breathed new life into* that idea.

IDEAS ARE PLANTS

His ideas have finally come to *fruition*. That idea *died on the vine*. That's a *budding* theory. It will take years for that idea to *come to full flower*. He views chemistry as a mere *offshoot* of physics. Mathematics has many *branches*. The *seeds* of his great ideas were *planted* in his youth. She has a *fertile* imagination. Here's an idea that I'd like to *plant* in your mind. He has a *barren* mind.

IDEAS ARE PRODUCTS

We're really *turning* (*churning, cranking, grinding*) *out* new ideas. We've *generated* a lot of ideas this week. He *produces* new ideas at an astounding rate. His *intellectual productivity* has decreased in recent years. We need to *take the rough edges off* that idea, *hone it down, smooth it out*. It's a rough idea; it needs to be *refined*.

IDEAS ARE COMMODITIES

It's important how you *package* your ideas. He won't *buy* that. That idea just won't *sell*. There is always a *market* for good ideas. That's a *worthless* idea. He's been a source of *valuable* ideas. I wouldn't *give a plugged nickel for* that idea. Your ideas don't have a chance in the *intellectual marketplace*.

IDEAS ARE RESOURCES

He *ran out of* ideas. Don't *waste* your thoughts on small projects. Let's *pool* our ideas. He's a *resourceful* man. We've *used up* all our ideas. That's a *useless* idea. That idea will *go a long way*.

IDEAS ARE MONEY

Let me put in my *two cents' worth*. He's *rich* in ideas. That book is a *treasure trove* of ideas. He has a *wealth* of ideas.

IDEAS ARE CUTTING INSTRUMENTS

That's an *incisive* idea. That *cuts right to the heart of* the matter. That was a *cutting* remark. He's *sharp*. He has a *razor* wit. He has a *keen* mind. She *cut* his argument *to ribbons*.

IDEAS ARE FASHIONS

That idea went *out of style* years ago. I hear sociobiology *is in* these days. Marxism is currently *fashionable* in western Europe. That idea is *old hat!* That's an *outdated* idea. What are the new *trends* in English criticism? *Old-fashioned* notions have no place in today's society. He keeps *up-to-date* by reading the New York Review of Books. Berkeley is a center of *avant-garde* thought. Semiotics has become quite *chic*. The idea of revolution is no longer *in vogue* in the United States. The transformational grammar *craze* hit the United States in the mid-sixties and has just made it to Europe.

UNDERSTANDING IS SEEING; IDEAS ARE LIGHT-SOURCES; DISCOURSE IS A LIGHT-MEDIUM

I *see* what you're saying. It *looks* different from my *point of view*. What is your *outlook* on that? I *view* it differently. Now I've got the *whole picture*. Let me *point something out* to you. That's an *insightful* idea. That was a *brilliant* remark. The argument is *clear*. It was a *murky* discussion. Could you *elucidate* your remarks? It's a *transparent* argument. The discussion was *opaque*.

LOVE IS A PHYSICAL FORCE (ELECTROMAGNETIC, GRAVITA-
TIONAL, etc.)

I could feel the *electricity* between us. There were *sparks*. I was *magnetically drawn* to her. They are uncontrollably *attracted* to each other. They *gravitated* to each other immediately. His whole life *revolves* around her. The *atmosphere* around them is always *charged*. There is incredible *energy* in their relationship. They lost their *momentum*.

LOVE IS A PATIENT

This is a *sick* relationship. They have a *strong, healthy* marriage. The marriage is *dead*—it can't be *revived*. Their marriage is *on the mend*. We're getting *back on our feet*. Their relationship is *in really good shape*. They've got a *listless* marriage. Their marriage is *on its last legs*. It's a *tired* affair.

LOVE IS MADNESS

I'm *crazy* about her. She *drives me out of my mind*. He constantly *raves* about her. He's gone *mad* over her. I'm just *wild* about Harry. I'm *insane* about her.

LOVE IS MAGIC

She *cast her spell* over me. The *magic* is gone. I was *spellbound*. She had me *hypnotized*. He has me *in a trance*. I was *entranced* by him. I'm *charmed* by her. She is *bewitching*.

LOVE IS WAR

He is known for his many rapid *conquests*. She *fought for* him, but his mistress *won out*. He *fled from* her *advances*. She *pursued* him *relentlessly*. He is slowly *gaining ground* with her. He *won* her hand in marriage. He *overpowered* her. She is *besieged* by suitors. He has to *fend* them *off*. He *enlisted the aid* of her friends. He *made an ally* of her mother. Theirs is a *misalliance* if I've ever seen one.

WEALTH IS A HIDDEN OBJECT

He's *seeking* his fortune. He's flaunting his *new-found* wealth. He's a *fortune-hunter*. She's a *gold-digger*. He *lost* his fortune. He's *searching for* wealth.

SIGNIFICANT IS BIG

He's a *big* man in the garment industry. He's a *giant* among writers. That's the *biggest* idea to hit advertising in years. He's *head and shoulders above* everyone in the industry. It was only a *small* crime. That was only a *little* white lie. I was astounded at the *enormity* of the crime. That was one of the *greatest* moments in World Series history. His accomplishments *tower over* those of *lesser* men.

SEEING IS TOUCHING; EYES ARE LIMBS

I can't *take* my eyes *off* her. He sits with his eyes *glued to* the TV. Her eyes *picked out* every detail of the pattern. Their eyes *met*. She never *moves* her eyes *from* his face. She *ran* her eyes *over* everything in the room. He wants everything *within reach of* his eyes.

THE EYES ARE CONTAINERS FOR THE EMOTIONS

I could see the fear *in* his eyes. His eyes were *filled* with anger. There was passion *in* her eyes. His eyes *displayed* his compassion. She couldn't *get* the fear *out* of her eyes. Love *showed in* his eyes. Her eyes *welled* with emotion.

EMOTIONAL EFFECT IS PHYSICAL CONTACT

His mother's death *hit* him *hard*. That idea *bowled me over*. She's a *knockout*. I was *struck* by his sincerity. That really *made an impression* on me. He *made his mark on* the world. I was *touched* by his remark. That *blew me away*.

PHYSICAL AND EMOTIONAL STATES ARE ENTITIES WITHIN A PERSON

He has a pain *in* his shoulder. Don't *give* me the flu. My cold has *gone from my head to my chest*. His pains *went away*. His depression *returned*. Hot tea and honey will *get rid of* your cough. He could barely *contain* his joy. The smile *left* his face. *Wipe* that sneer *off* your face, private! His fears *keep coming back*. I've got to *shake off* this depression—it keeps *hanging on*. If you've got a cold, drinking lots of tea will *flush it out* of your system. There isn't a *trace* of cowardice *in* him. He hasn't got *an honest bone in his body*.

VITALITY IS A SUBSTANCE

She's *brimming* with vim and vigor. She's *overflowing* with vitality. He's *devoid* of energy. I don't *have* any energy *left* at the end of the day. I'm *drained*. That *took a lot out of* me.

LIFE IS A CONTAINER

I've had a *full* life. Life is *empty* for him. There's *not much left* for him *in* life. Her life is *crammed* with activities. *Get the most out of* life. His life *contained* a great deal of sorrow. Live your life *to the fullest*.

LIFE IS A GAMBLING GAME

I'll *take my chances*. The *odds are against me*. I've got an *ace up my sleeve*. He's *holding all the aces*. It's a *toss-up*. If you *play your cards right*, you can do it. He *won big*. He's a real *loser*. Where is he when the *chips are down?* That's my *ace in the hole*. He's *bluffing*. The president is *playing it close to his vest*. *Let's up the ante*. Maybe we need to *sweeten the pot*. I think we should *stand pat*. That's *the luck of the draw*. Those are *high stakes*.

In this last group of examples we have a collection of what are called "speech formulas," or "fixed-form expressions," or "phrasal lexical items." These function in many ways like single words, and the language has thousands of them. In the examples given, a set of such phrasal lexical items is coherently structured by a single metaphorical concept. Although each of them is an instance of the LIFE IS A GAMBLING GAME metaphor, they are typically used to speak of life, not of gambling situations. They are normal ways of talking about life situations, just as using the word "construct" is a normal way of talking about theories. It is in this sense that we include them in what we have called literal expressions structured by metaphorical concepts. If you say "The odds are against us" or "We'll have to take our chances," you would not be viewed as speaking metaphorically but as using the normal everyday language appropriate to the situation. Nevertheless, your way of talking about, conceiving, and even experiencing your situation would be metaphorically structured.

11

The Partial Nature of Metaphorical Structuring

Up to this point we have described the systematic character of metaphorically defined concepts. Such concepts are understood in terms of a number of different metaphors (e.g., TIME IS MONEY, TIME IS A MOVING OBJECT, etc.). The metaphorical structuring of concepts is necessarily partial and is reflected in the lexicon of the language, including the phrasal lexicon, which contains fixed-form expressions such as "to be without foundation." Because concepts are metaphorically structured in a systematic way, e.g., THEORIES ARE BUILDINGS, it is possible for us to use expressions (*construct*, *foundation*) from one domain (BUILDINGS) to talk about corresponding concepts in the metaphorically defined domain (THEORIES). What *foundation*, for example, means in the metaphorically defined domain (THEORY) will depend on the details of how the metaphorical concept THEORIES ARE BUILDINGS is used to structure the concept THEORY.

The parts of the concept BUILDING that are used to structure the concept THEORY are the foundation and the outer shell. The roof, internal rooms, staircases, and hallways are parts of a building not used as part of the concept THEORY. Thus the metaphor THEORIES ARE BUILDINGS has a "used" part (foundation and outer shell) and an "unused" part (rooms, staircases, etc.). Expressions such as *construct* and *foundation* are instances of the used part of such a metaphorical concept and are part of our ordinary literal language about theories.

But what of the linguistic expressions that reflect the

"unused" part of a metaphor like THEORIES ARE BUILD-
INGS? Here are four examples:

His theory has thousands of little rooms and long, winding
corridors.
His theories are Bauhaus in their pseudofunctional sim-
plicity.
He prefers massive Gothic theories covered with gargoyles.
Complex theories usually have problems with the plumbing.

These sentences fall outside the domain of normal literal
language and are part of what is usually called "figurative"
or "imaginative" language. Thus, literal expressions ("He
has constructed a theory") and imaginative expressions
("His theory is covered with gargoyles") can be instances
of the same general metaphor (THEORIES ARE BUILDINGS).

Here we can distinguish three different subspecies of
imaginative (or nonliteral) metaphor:

Extensions of the used part of a metaphor, e.g., "These facts
are the bricks and mortar of my theory." Here the outer shell of
the building is referred to, whereas the THEORIES ARE BUILD-
INGS metaphor stops short of mentioning the materials used.

Instances of the unused part of the literal metaphor, e.g., "His
theory has thousands of little rooms and long, winding corridors."

Instances of novel metaphor, that is, a metaphor not used to
structure part of our normal conceptual system but as a new
way of thinking about something, e.g., "Classical theories are
patriarchs who father many children, most of whom fight in-
cessantly." Each of these subspecies lies outside the *used* part
of a metaphorical concept that structures our normal conceptual system.

We note in passing that all of the linguistic expressions
we have given to characterize general metaphorical concepts are figurative. Examples are TIME IS MONEY, TIME IS
A MOVING OBJECT, CONTROL IS UP, IDEAS ARE FOOD,
THEORIES ARE BUILDINGS, etc. None of these is literal. This

is a consequence of the fact that only *part* of them is used to structure our normal concepts. Since they necessarily contain parts that are not used in our normal concepts, they go beyond the realm of the literal.

Each of the metaphorical expressions we have talked about so far (e.g., the time *will come;* we *construct* a theory, *attack* an idea) is used within a whole system of metaphorical concepts—concepts that we constantly use in living and thinking. These expressions, like all other words and phrasal lexical items in the language, are fixed by convention. In addition to these cases, which are parts of whole metaphorical systems, there are idiosyncratic metaphorical expressions that stand alone and are not used systematically in our language or thought. These are well-known expressions like the *foot* of the mountain, a *head* of cabbage, the *leg* of a table, etc. These expressions are isolated instances of metaphorical concepts, where there is only one instance of a used part (or maybe two or three). Thus the *foot* of the mountain is the only used part of the metaphor A MOUNTAIN IS A PERSON. In normal discourse we do not speak of the *head, shoulders,* or *trunk* of a mountain, though in special contexts it is possible to construct novel metaphorical expressions based on these unused parts. In fact, there is an aspect of the metaphor A MOUNTAIN IS A PERSON in which mountain climbers will speak of the *shoulder* of a mountain (namely, a ridge near the top) and of *conquering, fighting,* and even *being killed by* a mountain. And there are cartoon conventions where mountains become animate and their peaks become heads. The point here is that there are metaphors, like A MOUNTAIN IS A PERSON, that are marginal in our culture and our language; their used part may consist of only one conventionally fixed expression of the language, and they do not systematically interact with other metaphorical concepts because so little of them is used. This makes them relatively uninteresting for our purposes but not completely so, since they can be extended to their unused part in coining novel metaphorical expressions, making jokes, etc. And our ability to extend them to unused parts indicates that, however marginal they are, they do exist.

Examples like the *foot* of the mountain are idiosyncratic, unsystematic, and isolated. They do not interact with other metaphors, play no particularly interesting role in our conceptual system, and hence are not metaphors that we live by. The only signs of life they have is that they can be extended in subcultures and that their unused portions serve as the basis for (relatively uninteresting) novel metaphors. If any metaphorical expressions deserve to be called "dead," it is these, though they do have a bare spark of life, in that they are understood partly in terms of marginal metaphorical concepts like A MOUNTAIN IS A PERSON.

It is important to distinguish these isolated and unsystematic cases from the systematic metaphorical expressions we have been discussing. Expressions like *wasting time, attacking positions, going our separate ways,* etc., are reflections of systematic metaphorical concepts that structure our actions and thoughts. They are "alive" in the most fundamental sense: they are metaphors we live by. The fact that they are conventionally fixed within the lexicon of English makes them no less alive.

12

How Is Our Conceptual System Grounded?

We claim that most of our normal conceptual system is metaphorically structured; that is, most concepts are partially understood in terms of other concepts. This raises an important question about the grounding of our conceptual system. Are there any concepts at all that are understood directly, without metaphor? If not, how can we understand anything at all?

The prime candidates for concepts that are understood directly are the simple spatial concepts, such as UP. Our spatial concept UP arises out of our spatial experience. We have bodies and we stand erect. Almost every movement we make involves a motor program that either changes our up-down orientation, maintains it, presupposes it, or takes it into account in some way. Our constant physical activity in the world, even when we sleep, makes an up-down orientation not merely relevant to our physical activity but centrally relevant. The centrality of up-down orientation in our motor programs and everyday functioning might make one think that there could be no alternative to this orientational concept. Objectively speaking, however, there are many possible frameworks for spatial orientation, including Cartesian coordinates, that don't in themselves have up-down orientation. Human spatial concepts, however, include UP-DOWN, FRONT-BACK, IN-OUT, NEAR-FAR, etc. It is these that are relevant to our continual everyday bodily functioning, and this gives them priority over other possible structurings of space—for us. In other words, the structure of our spatial concepts emerges from our constant spatial experience, that is, our interaction with the physical environment. Concepts that emerge in this way are concepts that we live by in the most fundamental way.

Thus UP is not understood purely in its own terms but emerges from the collection of constantly performed motor functions having to do with our erect position relative to the gravitational field we live in. Imagine a spherical being living outside any gravitational field, with no knowledge or imagination of any other kind of experience. What could UP possibly mean to such a being? The answer to this question would depend, not only on the physiology of this spherical being, but also on its culture.

In other words, what we call "direct physical experience" is never merely a matter of having a body of a certain sort; rather, *every* experience takes place within a vast background of cultural presuppositions. It can be misleading, therefore, to speak of direct physical experience as though there were some core of immediate experience which we then "interpret" in terms of our conceptual system. Cultural assumptions, values, and attitudes are not a conceptual overlay which we may or may not place upon experience as we choose. It would be more correct to say that all experience is cultural through and through, that we experience our "world" in such a way that our culture is already present in the very experience itself.

However, even if we grant that every experience involves cultural presuppositions, we can still make the important distinction between experiences that are "more" physical, such as standing up, and those that are "more" cultural, such as participating in a wedding ceremony. When we speak of "physical" versus "cultural" experience in what follows, it is in this sense that we use the terms.

Some of the central concepts in terms of which our bodies function—UP-DOWN, IN-OUT, FRONT-BACK, LIGHT-DARK, WARM-COLD, MALE-FEMALE, etc.—are more sharply delineated than others. While our emotional experience is as basic as our spatial and perceptual experience, our emotional experiences are much less sharply delineated in terms of what we do with our bodies. Although a sharply delineated conceptual structure for space emerges from our perceptual-motor functioning, no sharply defined conceptual structure for the emotions emerges from our emotional functioning alone. Since there are *systematic correlates* between our emotions (like happiness) and our sensory-motor experiences (like erect posture), these form the basis of orientational metaphorical concepts (such as HAPPY IS UP). Such metaphors allow us to conceptualize our emotions in more sharply defined terms and also to relate them to other concepts having to do with general well-being (e.g., HEALTH, LIFE, CONTROL, etc.). In this sense, we can speak of *emergent metaphors* and *emergent concepts.*

For example, the concepts OBJECT, SUBSTANCE, and CONTAINER emerge directly. We experience ourselves as entities, separate from the rest of the world—as containers with an inside and an outside. We also experience things external to us as entities—often also as containers with insides and outsides. We experience ourselves as being made up of substances—e.g., flesh and bone—and external objects as being made up of various kinds of substances— wood, stone, metal, etc. We experience many things, through sight and touch, as having distinct boundaries, and, when things have no distinct boundaries, we often project boundaries upon them—conceptualizing them as entities and often as containers (for example, forests, clearings, clouds, etc.).

As in the case of orientational metaphors, basic ontological metaphors are grounded by virtue of *systematic correlates within our experience.* As we saw, for example, the metaphor THE VISUAL FIELD IS A CONTAINER is grounded in the correlation between what we see and a bounded physical space. The TIME IS A MOVING OBJECT metaphor is based on the correlation between an object moving toward us and the time it takes to get to us. The same correlation is a basis for the TIME IS A CONTAINER metaphor (as in "He did it *in* ten minutes"), with the bounded space traversed by the object correlated with the time the object takes to traverse it. Events and actions are correlated with bounded time spans, and this makes them CONTAINER OBJECTS.

Experience with physical objects provides the basis for metonymy. Metonymic concepts emerge from correlations in our experience between two physical entities (e.g., PART FOR WHOLE, OBJECT FOR USER) or between a physical entity and something metaphorically conceptualized as a physical entity (e.g., THE PLACE FOR THE EVENT, THE INSTITUTION FOR THE PERSON RESPONSIBLE).

Perhaps the most important thing to stress about grounding is the distinction between an experience and the way we conceptualize it. We are not claiming that physical experience is in any way more basic than other kinds of experience, whether emotional, mental, cultural, or whatever. All of these experiences may be just as basic as physical experiences. Rather, what we are claiming about grounding is that we typically conceptualize the nonphysical *in terms of* the physical—that is, we conceptualize the less clearly delineated in terms of the more clearly delineated. Consider the following examples:

> Harry is in the kitchen.
> Harry is in the Elks.
> Harry is in love.

The sentences refer to three different domains of experience: spatial, social, and emotional. None of these has experiential priority over the others; they are all equally basic kinds of experience.

But with respect to conceptual structuring there is a difference. The concept IN of the first sentence emerges directly from spatial experience in a clearly delineated fashion. It is not an instance of a metaphorical concept. The other two sentences, however, are instances of metaphorical concepts. The second is an instance of the SOCIAL GROUPS ARE CONTAINERS metaphor, in terms of which the concept of a social group is structured. This metaphor allows us to "get a handle on" the concept of a social group by means of a spatialization. The word "in" and the concept IN are the same in all three examples; we do not have three different concepts of IN or three homophonous words "in." We have one emergent concept IN, one word for it, and two metaphorical concepts that partially define social groups and emotional states. What these cases show is that it is possible to have equally basic kinds of experiences while having conceptualizations of them that are not equally basic.

13

The Grounding of Structural Metaphors

Metaphors based on simple physical concepts—up-down, in-out, object, substance, etc.—which are as basic as anything in our conceptual system and without which we could not function in the world—could not reason or communicate—are not in themselves very rich. To say that something is viewed as a CONTAINER OBJECT with an IN-OUT orientation does not say very much about it. But, as we saw with the MIND IS A MACHINE metaphor and the various personification metaphors, we can elaborate spatialization metaphors in much more specific terms. This allows us not only to elaborate a concept (like the MIND) in considerable detail but also to find appropriate means for highlighting some aspects of it and hiding others. Structural metaphors (such as RATIONAL ARGUMENT IS WAR) provide the richest source of such elaboration. Structural metaphors allow us to do much more than just orient concepts, refer to them, quantify them, etc., as we do with simple orientational and ontological metaphors; they allow us, in addition, to use one highly structured and clearly delineated concept to structure another.

Like orientational and ontological metaphors, structural metaphors are grounded in systematic correlations within our experience. To see what this means in detail, let us examine how the RATIONAL ARGUMENT IS WAR metaphor might be grounded. This metaphor allows us to conceptualize what a rational argument is in terms of something that we understand more readily, namely, physical conflict. Fighting is found everywhere in the animal kingdom and nowhere so much as among human animals. Animals fight to get what they want—food, sex, territory, control, etc.—because there are other animals who want the same thing or who want to stop them from getting it. The same is true of human animals, except that we have developed more sophisticated techniques for getting our way. Being "rational animals," we have institutionalized our fighting in a number of ways, one of them being war. Even though we have over the ages institutionalized physical conflict and have employed many of our finest minds to develop more effective means of carrying it out, its basic structure remains essentially unchanged. In fights between two brute animals, scientists have observed the practices of issuing challenges for the sake of intimidation, of establishing and defending territory, attacking, defending, counterattacking, retreating, and surrendering. Human fighting involves the same practices.

Part of being a rational animal, however, involves getting what you want without subjecting yourself to the dangers of actual physical conflict. As a result, we humans have evolved the social institution of verbal argument. We have arguments all the time in order to try to get what we want, and sometimes these "degenerate" into physical violence. Such verbal battles are comprehended in much the same terms as physical battles. Take a domestic quarrel, for instance. Husband and wife are both trying to get what each of them wants, such as getting the other to accept a certain viewpoint on some issue or at least to act according to that viewpoint. Each sees himself as having something to win and something to lose, territory to establish and territory to defend. In a no-holds-barred argument, you attack, defend, counterattack, etc., using whatever verbal means you have at your disposal—intimidation, threat, invoking authority, insult, belittling, challenging authority, evading issues, bargaining, flattering, and even trying to give "rational reasons." But all of these tactics can be, and often are, presented as *reasons;* for example:

... because I'm bigger than you. (*intimidation*)
... because if you don't, I'll... (*threat*)
... because I'm the boss. (*authority*)
... because you're stupid. (*insult*)
... because you usually do it wrong. (*belittling*)
... because I have as much right as you do. (*challenging authority*)
... because I love you. (*evading the issue*)
... because if you will..., I'll... (*bargaining*)
... because you're so much better at it. (*flattery*)

Arguments that use tactics like these are the most common in our culture, and because they are so much a part of our daily lives, we sometimes don't notice them. However, there are important and powerful segments of our culture where such tactics are, at least in principle, frowned upon because they are considered to be "irrational" and "unfair." The academic world, the legal world, the diplomatic world, the ecclesiastical world, and the world of journalism claim to present an ideal, or "higher," form of RATIONAL ARGUMENT, in which all of these tactics are forbidden. The only permissible tactics in this RATIONAL ARGUMENT are supposedly the stating of premises, the citing of supporting evidence, and the drawing of logical conclusions. But even in the most ideal cases, where all of these conditions hold, RATIONAL ARGUMENT is still comprehended and carried out in terms of WAR. There is still a position to be established and defended, you can win or lose, you have an opponent whose position you attack and try to destroy and whose argument you try to shoot down. If you are completely successful, you can wipe him out.

The point here is that not only our conception of an argument but the way we carry it out is grounded in our knowledge and experience of physical combat. Even if you have never fought a fistfight in you life, much less a war, but have been arguing from the time you began to talk, you still conceive of arguments, and execute them, according to the

ARGUMENT IS WAR metaphor because the metaphor is built into the conceptual system of the culture in which you live. Not only are all the "rational" arguments that are assumed to actually live up to the ideal of RATIONAL ARGUMENT conceived of in terms of WAR, but almost all of them contain, in hidden form, the "irrational" and "unfair" tactics that rational arguments in their ideal form are supposed to transcend. Here are some typical examples:

It is plausible to assume that... (*intimidation*)
Clearly,...
Obviously,...

It would be unscientific to fail to... (*threat*)
To say that would be to commit the Fallacy of...

As Descartes showed,... (*authority*)
Hume observed that...
Footnote 374: cf. Verschlugenheimer, 1954.

The work lacks the necessary rigor for... (*insult*)
Let us call such a theory "Narrow" Rationalism.
In a display of "scholarly objectivity,"...

The work will not lead to a formalized theory. (*belittling*)
His results cannot be quantified.
Few people today seriously hold that view.

Lest we succumb to the error of positivist approaches,...
    (*challenging authority*)
Behaviorism has led to...

He does not present any alternative theory. (*evading the issue*)
But that is a matter of...
The author does present some challenging facts, although...

Your position is right as far as it goes,... (*bargaining*)
If one takes a realist point of view, one can accept the claim
    that...

In his stimulating paper,... (*flattery*)
His paper raises some interesting issues...

Examples like these allow us to trace the lineage of our rational argument back through "irrational" argument (= *everyday arguing*) to its origins in physical combat. The tactics of intimidation, threat, appeal to authority, etc., though couched, perhaps, in more refined phrases, are just as present in rational argument as they are in everyday arguing and in war. Whether we are in a scientific, academic, or legal setting, aspiring to the ideal of rational argument, or whether we are just trying to get our way in our own household by haggling, the way we conceive of, carry out, and describe our arguments is grounded in the ARGUMENT IS WAR metaphor.

Let us now consider other structural metaphors that are important in our lives: LABOR IS A RESOURCE and TIME IS A RESOURCE. Both of these metaphors are culturally grounded in our experience with material resources. Material resources are typically raw materials or sources of fuel. Both are viewed as serving purposeful ends. Fuel may be used for heating, transportation, or the energy used in producing a finished product. Raw materials typically go directly into products. In both cases, the material resources can be *quantified* and given a *value*. In both cases, it is the *kind* of material as opposed to the particular piece or quantity of it that is important for achieving the purpose. For example, it doesn't matter which particular pieces of coal heat your house as long as they are the right *kind* of coal. In both cases, the material gets *used up* progressively as the purpose is served. To summarize:

A material resource is a *kind* of substance
    can be *quantified* fairly precisely
    can be assigned a *value* per unit quantity
    serves a *purposeful* end
    is *used up* progressively as it serves its
    purpose

Take the simple case where you make a product from raw material. It takes a certain amount of labor. In general, the more labor you perform, the more you produce. Assuming that this is true—that the labor is proportional to the amount of product—we can assign *value* to the labor in terms of the time it takes to produce a unit of product. The perfect model of this is the assembly line, where the raw material comes in at one end, labor is performed in progressive stages, whose duration is fixed by the speed of the line itself, and products come out at the other end. This provides a grounding for the LABOR IS RESOURCE metaphor, as follows:

> LABOR is a *kind* of activity (recall: AN ACTIVITY IS A SUBSTANCE)
>> can be *quantified* fairly precisely (in terms of time)
>> can be assigned a *value* per unit
>> serves a *purposeful* end
>> is *used up* progressively as it serves it purpose

Since labor can be quantified in terms of time and usually is, in an industrial society, we get the basis for the TIME IS A RESOURCE metaphor:

> TIME is a *kind* of (abstract) SUBSTANCE
>> can be *quantified* fairly precisely
>> can be assigned a *value* per unit
>> serves a *purposeful* end
>> is *used up* progressively as it serves its purpose

When we are living by the metaphors LABOR IS A RESOURCE and TIME IS A RESOURCE, as we do in our culture, we tend not to see them as metaphors at all. But, as the above account of their grounding in experience shows, both are structural metaphors that are basic to Western industrial societies.

These two complex structural metaphors both employ simple ontological metaphors. LABOR IS A RESOURCE uses AN ACTIVITY IS A SUBSTANCE. TIME IS A RESOURCE uses TIME IS A SUBSTANCE. These two SUBSTANCE metaphors permit labor and time to be quantified—that is, measured, conceived of as being progressively "used up," and assigned monetary values; they also allow us to view time and labor as things that can be "used" for various ends.

LABOR IS A RESOURCE and TIME IS A RESOURCE are by no means universal. They emerged naturally in our culture because of the way we view work, our passion for quantification, and our obsession with purposeful ends. These metaphors highlight those aspects of labor and time that are centrally important in our culture. In doing this, they also deemphasize or hide certain aspects of labor and time. We can see what both metaphors hide by examining what they focus on.

In viewing labor as a *kind* of activity, the metaphor assumes that labor can be clearly identified and distinguished from things that are not labor. It makes the assumptions that we can tell work from play and productive activity from nonproductive activity. These assumptions obviously fail to fit reality much of the time, except perhaps on assembly lines, chain gangs, etc. The view of labor as merely a *kind* of activity, independent of who performs it, how he experiences it, and what it means in his life, hides the issues of whether the work is personally meaningful, satisfying, and humane.

The quantification of labor in terms of time, together with the view of time as serving a purposeful end, induces a notion of LEISURE TIME, which is parallel to the concept LABOR TIME. In a society like ours, where inactivity is not considered a purposeful end, a whole industry devoted to leisure activity has evolved. As a result, LEISURE TIME becomes a RESOURCE too—to be spent productively, used wisely, saved up, budgeted, wasted, lost, etc. What is hidden by the RESOURCE metaphors for labor and time is the way our concepts of LABOR and TIME affect our concept of LEISURE, turning it into something remarkably like LABOR.

The RESOURCE metaphors for labor and time hide all sorts of possible conceptions of labor and time that exist in other cultures and in some subcultures of our own society: the idea that work can be play, that inactivity can be productive, that much of what we classify as LABOR serves either no clear purpose or no worthwhile purpose.

The three structural metaphors we have considered in

this section—RATIONAL ARGUMENT IS WAR, LABOR IS A RE-
SOURCE, and TIME IS A RESOURCE—all have a strong cultural basis. They emerged naturally in a culture like ours
because what they highlight corresponds so closely to what
we experience collectively and what they hide corresponds
to so little. But not only are they grounded in our physical
and cultural experience; they also influence our experience
and our actions.

14

Causation: Partly Emergent and Partly Metaphorical

We have seen in our discussion of grounding that there are directly emergent concepts (like UP-DOWN, IN-OUT, OBJECT, SUBSTANCE, etc.) and emergent metaphorical concepts based on our experience (like THE VISUAL FIELD IS A CONTAINER, AN ACTIVITY IS A CONTAINER, etc.). From the limited range of examples we have considered, it might seem as if there were a clear distinction between directly emergent and metaphorically emergent concepts and that every concept must be one or the other. This is not the case. Even a concept as basic as CAUSATION is not purely emergent or purely metaphorical. Rather, it appears to have a directly emergent core that is elaborated metaphorically.

Direct Manipulation: The Prototype of Causation

Standard theories of meaning assume that all of our complex concepts can be analyzed into undecomposable primitives. Such primitives are taken to be the ultimate "building blocks" of meaning. The concept of causation is often taken to be such an ultimate building block. We believe that the standard theories are fundamentally mistaken in assuming that basic concepts are undecomposable primitives.

We agree that causation is a basic human concept. It is one of the concepts most often used by people to organize their physical and cultural realities. But this does not mean that it is an undecomposable primitive. We would like to suggest instead that causation is best understood as an experiential gestalt. A proper understanding of causation requires that it be viewed as a cluster of other components. But the cluster forms a gestalt—a whole that we human beings find more basic than the parts.

We can see this most clearly in infants. Piaget has hypothesized that infants first learn about causation by realizing that they can directly manipulate objects around them—pull off their blankets, throw their bottles, drop toys. There is, in fact, a stage in which infants seem to "practice" these manipulations, e.g., they repeatedly drop their spoons. Such direct manipulations, even on the part of infants, involve certain shared features that characterize the notion of direct causation that is so integral a part of our constant everyday functioning in our environment—as when we flip light switches, button our shirts, open doors, etc. Though each of these actions is different, the overwhelming proportion of them share features of what we may call a "prototypical" or "paradigmatic" case of direct causation. These shared features include:

The agent has as a goal some change of state in the patient.

The change of state is physical.

The agent has a "plan" for carrying out this goal.

The plan requires the agent's use of a motor program.

The agent is in control of that motor program.

The agent is primarily responsible for carrying out the plan.

The agent is the energy source (i.e., the agent is directing his energies toward the patient), and the patient is the energy goal (i.e., the change in the patient is due to an external source of energy).

The agent touches the patient either with his body or an instrument (i.e., there is a spatiotemporal overlap between what the agent does and the change in the patient).

The agent successfully carries out the plan.

The change in the patient is perceptible.

The agent monitors the change in the patient through sensory perception.

There is a single specific agent and a single specific patient.

This set of properties characterizes "prototypical" direct manipulations, and these are cases of causation par excellence. We are using the word "prototypical" in the sense Rosch uses it in her theory of human categorization (1977). Her experiments indicate that people categorize objects, not in set-theoretical terms, but in terms of prototypes and family resemblances. For example, small flying singing birds, like sparrows, robins, etc., are *prototypical birds*. Chickens, ostriches, and penguins are birds but are not central members of the category—they are nonprototypical birds. But they are birds nonetheless, because they bear sufficient family resemblances to the prototype; that is, they share enough of the relevant properties of the prototype to be classified by people as birds.

The twelve properties given above characterize a prototype of causation in the following sense. They recur together over and over in action after action as we go through our daily lives. We experience them as a *gestalt;* that is, the complex of properties occurring together is more basic to our experience than their separate occurrence. Through their constant recurrence in our everyday functioning, the category of causation emerges with this complex of properties characterizing prototypical causations. Other kinds of causation, which are less prototypical, are actions or events that bear sufficient family resemblances to the prototype. These would include action at a distance, nonhuman agency, the use of an intermediate agent, the occurrence of two or more agents, involuntary or uncontrolled use of the motor program, etc. (In physical causation the agent and patient are events, a physical law takes the place of plan, goal, and motor activity, and all of the peculiarly human aspects are factored out.) When there is an insufficient family resemblance to the prototype, we cease to characterize what happens as causation. For example, if there were multiple agents, if what the agents did was remote in space and time from the patient's change, and if there were neither desire nor plan nor control, then we probably wouldn't say that this was an instance of causation, or at least we would have questions about it.

Although the category of causation has fuzzy boundaries, it is clearly delineated in an enormous range of instances. Our successful functioning in the world involves the application of the concept of causation to ever new domains of activity—through intention, planning, drawing inferences, etc. The concept is stable because we continue to function successfully in terms of it. Given a concept of causation that emerges from our experience, we can apply that concept to metaphorical concepts. In "Harry raised our morale by telling jokes," for example, we have an instance of causation where what Harry did made our morale go UP, as in the HAPPY IS UP metaphor.

Though the concept of causation as we have characterized it is basic to human activity, it is not a "primitive" in the usual building-block sense, that is, it is not unanalyzable and undecomposable. Since it is defined in terms of a prototype that is characterized by a recurrent complex of properties, our concept of causation is at once holistic, analyzable into those properties, and capable of a wide range of variation. The terms into which the causation prototype is analyzed (e.g., control, motor program, volition, etc.) are probably also characterized by prototype and capable of further analysis. This permits us to have concepts that are at once basic, holistic, and indefinitely analyzable.

Metaphorical Extensions of Prototypical Causation

Simple instances of making an object (e.g., a paper airplane, a snowball, a sand castle) are all special cases of direct causation. They all involve prototypical direct manipulation, with all of the properties listed above. But they have one additional characteristic that sets them apart as instances of *making:* As a result of the manipulation, we view the object as a different *kind* of thing. What was a sheet of paper is now a paper airplane. We categorize it differently—it has a different form and function. It is essentially this that sets instances of *making* apart from other kinds of direct manipulation. Even a simple change of state, like the change from water to ice, can be viewed as an instance of making, since ice has a different form and function than water. Thus we get examples like:

> You can make ice out of water by freezing it.

This parallels examples like:

> I made a paper airplane out of a sheet of newspaper.
> I made a statue out of clay.

We conceptualize changes of this kind—from one state into another, having a new form and function in terms of the metaphor THE OBJECT COMES OUT OF THE SUBSTANCE. This is why the expression *out of* is used in the above examples: the ice is viewed as emerging out of the water; the airplane is viewed as emerging out of the paper; the statue is viewed as emerging out of the clay. In a sentence like "I made a statue out of clay," the substance clay is viewed as the CONTAINER (via the SUBSTANCE IS A CONTAINER metaphor) from which the object—namely, the statue—emerges. Thus the concept MAKING is partly, but not totally, metaphorical. That is, MAKING is an instance of a directly emergent concept, namely, DIRECT MANIPULATION, which is further elaborated by the metaphor THE OBJECT COMES OUT OF THE SUBSTANCE.

Another way we can conceptualize making is by elaborating on direct manipulation, using another metaphor: THE SUBSTANCE GOES INTO THE OBJECT. Thus:

> I made a sheet of newspaper *into* an airplane.
> I made the clay you gave me *into* a statue.

Here the object is viewed as a container for the material. The SUBSTANCE GOES INTO THE OBJECT metaphor occurs far more widely than in the concept of MAKING. We conceptualize a wide range of changes, natural as well as man-made, in terms of this metaphor. For example:

The water turned *into* ice.
The caterpillar turned *into* a butterfly.
She is slowly changing *into* a beautiful woman.

The OBJECT COMES OUT OF THE SUBSTANCE metaphor is also used outside the concept of MAKING but in a much more limited range of circumstances, mostly those having to do with evolution:

Mammals developed *out of* reptiles.
Our present legal system evolved *out of* English common law.

Thus the two metaphors we use to elaborate direct manipulation into the concept of MAKING are both used independently to conceptualize various concepts of CHANGE.

These two metaphors for CHANGE, which are used as part of the concept of MAKING, emerge naturally from as fundamental a human experience as there is, namely, birth. In birth, an object (the baby) comes out of a container (the mother). At the same time, the mother's substance (her flesh and blood) are in the baby (the container object). The experience of birth (and also agricultural growth) provides a grounding for the general concept of CREATION, which has as its core the concept of MAKING a physical object but which extends to abstract entities as well. We can see this grounding in birth metaphors for creation in general:

Our nation was *born out of* a desire for freedom.
His writings are products of his *fertile* imagination.
His experiment *spawned* a host of new theories.
Your actions will only *breed* violence.
He *hatched* a clever scheme.
He *conceived* a brilliant theory of molecular motion.
Universities are *incubators* for new ideas.
The theory of relativity *first saw the light of day* in 1905.

The University of Chicago was the *birthplace* of the nuclear age.
Edward Teller is the *father* of the hydrogen bomb.

These are all instances of the general metaphor CREATION IS BIRTH. This gives us another instance where a special case of causation is conceptualized metaphorically.

Finally, there is another special case of CAUSATION which we conceptualize in terms of the EMERGENCE metaphor. This is the case where a mental or emotional state is viewed as causing an act or event:

He shot the mayor *out of* desperation.
He gave up his career *out of* love for his family.
His mother nearly went crazy *from* loneliness.
He dropped *from* exhaustion.
He became a mathematician *out of* a passion for order.

Here the STATE (desperation, loneliness, etc.) is viewed as a container, and the act or event is viewed as an object that emerges from the container. The CAUSATION is viewed as the EMERGENCE of the EVENT from the STATE.

Summary

As we have just seen, the concept of CAUSATION is based on the prototype of DIRECT MANIPULATION, which emerges directly from our experience. The prototypical core is elaborated by metaphor to yield a broad concept of CAUSATION, which has many special cases. The metaphors used are THE OBJECT COMES OUT OF THE SUBSTANCE, THE SUBSTANCE GOES INTO THE OBJECT, CREATION IS BIRTH, and CAUSATION (of event by state) IS EMERGENCE (of the event/object from the state/container).

We also saw that the prototypical core of the concept CAUSATION, namely, DIRECT MANIPULATION, is not an unanalyzable semantic primitive but rather a gestalt consisting of properties that naturally occur together in our daily experience of performing direct manipulations. The prototypical concept DIRECT MANIPULATION is basic and primitive in our experience, but not in the sense required by a "building-block" theory. In such theories, each concept either *is* an ultimate building block or can be broken down into ultimate building blocks in one and only one way. The theory we will propose in the next chapter suggests, instead, that there are natural dimensions of experience and that concepts can be analyzed along these dimensions in more than one way. Moreover, along each dimension, concepts can often be analyzed further and further, relative to our experience, so that there are not always ultimate building blocks.

Thus there are three ways in which CAUSATION is not an unanalyzable primitive:

—It is characterized in terms of family resemblances to the prototype of DIRECT MANIPULATION.

—The DIRECT MANIPULATION prototype itself is an indefinitely analyzable gestalt of naturally cooccurring properties.

—The prototypical core of CAUSATION is elaborated metaphorically in various ways.

15

The Coherent Structuring of
Experience

Experiential Gestalts and the Dimensions of
Experience

We have talked throughout of metaphorical concepts as
ways of partially structuring one experience in terms of
another. In order to see in detail what is involved in
metaphorical structuring, we must first have a clearer idea
of what it means for an experience or set of experiences to
be coherent by virtue of having a structure. For example,
we have suggested that an argument is a conversation that
is partially structured by the concept WAR (thus giving us
the ARGUMENT IS WAR metaphor). Suppose you are having
a conversation and you suddenly realize that it has turned
into an argument. What is it that makes a conversation an
argument, and what does that have to do with war? To see
the difference between a conversation and an argument, we
first have to see what it means to be engaged in a conversation.

The most basic kind of conversation involves two people
who are talking to each other. Typically, one of them ini-
tiates it and they take turns talking about some common
topic or set of topics. Maintaining the turn-taking and
keeping to the topic at hand (or shifting topics in a permissible fashion) takes a certain amount of cooperation. And
whatever other purposes a conversation may have for the
participants, conversations generally serve the purpose of
polite social interaction.

Even in as simple a case as a polite two-party conversation, several dimensions of structure can be seen:

Participants: The participants are of a certain natural kind, namely, people. Here they take the role of speakers. The conversation is defined by what the participants do, and the same participants play a role throughout the conversation.

Parts: The parts consist of a certain natural kind of activity, namely, talking. Each turn at talking is a part of the conversation as a whole, and these parts must be put together in a certain fashion for there to be a coherent conversation.

Stages: Conversations typically have a set of initial conditions and then pass through various stages, including at least a beginning, a central part, and an end. Thus there are certain things that are said in order to initiate a conversation ("Hello!", "How are you?", etc.), others that move it along to the central part, and still others that end it.

Linear sequence: The participants' turns at speaking are ordered in a linear sequence, with the general constraint that the speakers alternate. Certain overlappings are permitted, and there are lapses where one speaker doesn't take his turn and the other speaker continues. Without such constraints on linear sequencing of parts, you get a monologue or a jumble of words but no conversation.

Causation: The finish of one turn at talking is expected to result in the beginning of the next turn.

Purpose: Conversations may serve any number of purposes, but all typical conversations share the purpose of maintaining polite social interaction in a reasonably cooperative manner.

There are many details that could be added that characterize conversation more precisely, but these six dimensions of structure give the main outlines of what is common to typical conversations.

If you are engaged in a conversation (which has at least these six dimensions of structure) and you perceive it turning into an argument, what is it that you perceive over and above being in a conversation? The basic difference is a sense of being embattled. You realize that you have an opinion that matters to you and that the other person doesn't accept it. At least one participant wants the other to give up his opinion, and this creates a situation where there is something to be won or lost. You sense that you are in an argument when you find your own position under attack or when you feel a need to attack the other person's position. It becomes a full-fledged argument when both of you devote most of your conversational energy to trying to discredit the other person's position while maintaining your own. The argument remains a conversation, although the element of polite cooperation in maintaining the conversational structure may be strained if the argument becomes heated.

The sense of being embattled comes from experiencing yourself as being in a warlike situation even though it is not actual combat—since you are maintaining the amenities of conversation. You experience the other participant as an adversary, you attack his position, you try to defend your own, and you do what you can to make him give in. The structure of the conversation takes on aspects of the structure of a war, and you act accordingly. Your perceptions and actions correspond in part to the perceptions and actions of a party engaged in war. We can see this in more detail in the following list of characteristics of argument:

> You have an opinion that matters to you. (*having a position*)
>
> The other participant does not agree with your opinion. (*has a different position*)
>
> It matters to one or both of you that the other give up his opinion (*surrender*) and accept yours (*victory*). (*he is your adversary*)
>
> The difference of opinion becomes a conflict of opinions. (*conflict*)
>
> You think of how you can best convince him of your view (*plan strategy*) and consider what evidence you can bring to bear on the issue (*marshal forces*).
>
> Considering what you perceive as the weaknesses of his position, you ask questions and raise objections designed to force him ultimately to give up his position and adopt yours. (*attack*)

You try to change the premises of the conversation so that you will be in a stronger position. (*maneuvering*)

In response to his questions and objections, you try to maintain your own position. (*defense*)

As the argument progresses, maintaining your general view may require some revision. (*retreat*)

You may raise new questions and objections. (*counterattack*)

Either you get tired and decide to quit arguing (*truce*), or neither of you can convince the other (*stalemate*), or one of you gives in. (*surrender*)

What gives coherence to this list of things that make a conversation into an argument is that they correspond to elements of the concept WAR. What is added from the concept WAR to the concept CONVERSATION can be viewed in terms of the same six dimensions of structure that we gave in our description of conversational structure.

Participants:	The kind of participants are people or groups of people. They play the role of adversaries.
Parts:	The two positions Planning strategy Attack Defense Retreat Maneuvering Counterattack Stalemate Truce Surrender/victory
Stages:	Initial conditions: Participants have different positions. One or both wants the other to surrender. Each participant assumes he can defend his position. Beginning: One adversary attacks. Middle: Combinations of defense maneuvering

retreat
counterattack
End: Either truce or stalemate or surrender/
victory
Final state: Peace, victor has dominance over
loser

Linear sequence: Retreat after attack
Defense after attack
Counterattack after attack

Causation: Attack results in defense or counterattack or
retreat or end.

Purpose: Victory

Understanding a conversation as being an argument involves being able to superimpose the multidimensional structure of part of the concept WAR upon the corresponding structure CONVERSATION. Such multidimensional structures characterize *experiential gestalts,* which are ways of organizing experiences into *structured wholes.* In the ARGUMENT IS WAR metaphor, the gestalt for CONVERSATION is structured further by means of correspondences with selected elements of the gestalt for WAR. Thus one activity, talking, is understood in terms of another, physical fighting. Structuring our experience in terms of such multidimensional gestalts is what makes our experience *coherent.* We experience a conversation as an argument when the WAR gestalt fits our perceptions and actions in the conversation.

Understanding such multidimensional gestalts and the correlations between them is the key to understanding coherence in our experience. As we saw above, *experiential gestalts are multidimensional structured wholes.* Their dimensions, in turn, are defined in terms of directly emergent concepts. That is, the various dimensions (participants, parts, stages, etc.) are categories that emerge naturally from our experience. We have already seen that CAUSATION is a directly emergent concept, and the other dimensions in terms of which we categorize our experience have a fairly obvious experiential basis:

Participants: This dimension arises out of the concept of the SELF as an actor distinguishable from the actions he performs. We also distinguish *kinds* of participants (e.g., people, animals, objects).

Parts: We experience ourselves as having parts (arms, legs, etc.) that we can control independently. Likewise, we experience physical objects either in terms of parts that they naturally have or parts that we impose upon them, either by virtue of our perceptions, our interactions with them, or our uses for them. Similarly, we impose a part-whole structure on events and activities. And, as in the case of participants, we distinguish *kinds* of parts (e.g., kinds of objects, kinds of activities, etc.).

Stages: Our simplest motor functions involve knowing where we are and what position we are in (initial conditions), starting to move (beginning), carrying out the motor function (middle), and stopping (end), which leaves us in a final state.

Linear sequence: Again, the control of our simplest motor functions requires us to put them in the right linear sequence.

Purpose: From birth (and even before), we have needs and desires, and we realize very early that we can perform certain actions (crying, moving, manipulating objects) to satisfy them.

These are some of the basic dimensions of our experience. We classify our experiences in such terms. And we see *coherence* in diverse experiences when we can categorize them in terms of gestalts with at least these dimensions.

What Does It Mean for a Concept to Fit an Experience?

Let us return to the experience of being in a conversation that turns into an argument. As we saw, being in a conversation is a structured experience. As we experience a conversation, we are automatically and unconsciously classifying our experience in terms of the natural dimensions of the CONVERSATION gestalt: Who's participating? Whose turn is it? (= which part?) What stage are we at? And so on. It is in terms of imposing the CONVERSATION gestalt on what is happening that we experience the talking and listening that we engage in as a particular *kind* of experience, namely, a conversation. When we perceive dimensions of our experience as fitting the WAR gestalt in addition, we become aware that we are participating in another *kind* of experience, namely, an argument. It is by this means that we classify particular experiences, and we need to classify our experiences in order to comprehend, so that we will know what to do.

Thus we classify particular experiences in terms of experiential gestalts in our conceptual system. Here we must distinguish between: (1) the experience itself, as we structure it, and (2) the concepts that we employ in structuring it, that is, the multidimensional gestalts like CONVERSATION and ARGUMENT. The concept (say, CONVERSATION) specifies certain natural dimensions (e.g., participants, parts, stages, etc.) and how these dimensions are related. There is a correlation, dimension by dimension, between the concept CONVERSATION and the aspects of the actual activity of conversing. This is what we mean when we say that a concept fits an experience.

It is by means of conceptualizing our experiences in this manner that we pick out the "important" aspects of an experience. And by picking out what is "important" in the experience, we can categorize the experience, understand it, and remember it. If we were to tell you that we had an argument yesterday, we would be telling you the truth if our concept of an ARGUMENT, with us as participants, fits an experience that we had yesterday, dimension by dimension.

Metaphorical Structuring versus Subcategorization

In our discussion of the concept ARGUMENT, we have been assuming a clear-cut distinction between subcategorization and metaphorical structuring. On the one hand, we took "An argument is a conversation" to be an instance of subcategorization, because an argument is basically a *kind* of conversation. The same kind of activity occurs in both, namely, talking, and an argument has all the basic structural features of a conversation. Thus our criteria for subcategorization were (*a*) same kind of activity and (*b*) enough of the same structural features. On the other hand, we took ARGUMENT IS WAR to be a metaphor because an argument and a war are basically different kinds of activity, and ARGUMENT is partially structured in terms of WAR. Argument is a different kind of activity because it involves talking instead of combat. The structure is partial, because only selected elements of the concept WAR are used. Thus our criteria for metaphor were (*a*) a difference in kind of activity and (*b*) partial structuring (use of certain selected parts).

But we cannot always distinguish subcategorization from metaphor on the basis of these criteria. The reason is that it is not always clear when two activities (or two things) are of the same kind or of different kinds. Take, for example, AN ARGUMENT IS A FIGHT. Is this a subcategorization or a metaphor? The issue here is whether fighting and arguing are the same kind of activity. This is not a simple issue. Fighting is an attempt to gain dominance that typically involves hurting, inflicting pain, injuring, etc. But there is both physical pain and what is called psychological pain; there is physical dominance and there is psychological dominance. If your concept FIGHT includes psychological dominance and psychological pain on a par with physical dominance and pain, then you may see AN ARGUMENT IS A FIGHT as a subcategorization rather than a metaphor, since both would involve gaining psychological dominance. On this view an argument would be a kind of fight, structured in the form of a conversation. If, on the other hand, you conceive of FIGHT as purely physical, and if you view psychological pain only as pain taken metaphorically, then you might view AN ARGUMENT IS A FIGHT as metaphorical.

The point here is that subcategorization and metaphor are endpoints on a continuum. A relationship of the form A is B (for example, AN ARGUMENT IS A FIGHT) will be a clear subcategorization if A and B are the same kind of thing or activity and will be a clear metaphor if they are clearly different kinds of things or activities. But when it is not clear whether A and B are the same kind of thing or activity, then the relationship A is B falls somewhere in the middle of the continuum.

The important thing to note is that the theory outlined in chapter 14 allows for such unclear cases as well as for the clear ones. The unclear cases will involve the same kinds of structures (with the same dimensions and the same possible complexities) as the clear cases. In an unclear case of the form A is B, A and B will both be gestalts that structure certain kinds of activities (or things), and the only question will be whether the activities or things structured by those gestalts are of the same *kind*.

We have so far characterized coherence in terms of experiential gestalts, which have various dimensions that emerge naturally from experience. Some gestalts are relatively simple (CONVERSATION) and some are extremely elaborate (WAR). There are also complex gestalts, which are structured partially in terms of other gestalts. These are what we have been calling *metaphorically structured concepts*. Certain concepts are structured almost entirely metaphorically. The concept LOVE, for example, is structured mostly in metaphorical terms: LOVE IS A JOURNEY, LOVE IS A PATIENT, LOVE IS A PHYSICAL FORCE, LOVE IS MADNESS, LOVE IS WAR, etc. The concept of LOVE has a core that is minimally structured by the subcategorization LOVE IS AN EMOTION and by links to other emotions, e.g., liking. This is typical of emotional concepts, which are not clearly delineated in our experience in any direct fashion and therefore must be comprehended primarily indirectly, via metaphor.

But there is more to coherence than structuring in terms

of multidimensional gestalts. When a concept is structured by more than one metaphor, the different metaphorical structurings usually fit together in a coherent fashion. We will now turn to other aspects of coherence, both within a single metaphorical structuring and across two or more metaphors.

16

Metaphorical Coherence

Specialized Aspects of a Concept

So far we have looked at the concept ARGUMENT in enough detail to get a sense of its general overall structure. As is the case with many of our general concepts, the concept AR-GUMENT has specialized aspects that are used in certain subcultures or in certain situations. We saw, for example, that in the academic world, legal world, etc., the concept ARGUMENT is specialized to RATIONAL ARGUMENT, which is distinguished from everyday, "irrational" argument. In RATIONAL ARGUMENT the tactics are *ideally* restricted to stating premises, citing supporting evidence, and drawing logical conclusions. In practice, as we saw, the tactics of everyday argument (intimidation, appeal to authority, etc.) appear in actual "rational" argument in a disguised or refined form. These additional restrictions define RATIONAL ARGUMENT as a specialized branch of the general concept ARGUMENT. Moreover, the purpose of argument is further restricted in the case of RATIONAL ARGUMENT. In the ideal case, the purpose of winning the argument is seen as serving the higher purpose of understanding.

Within RATIONAL ARGUMENT itself there is a further specialization. Since written discourse rules out the dialogue inherent in two-party arguments, a special form of one-party argument has developed. Here speaking typically becomes writing, and the author addresses himself, not to an actual adversary, but to a set of hypothetical adversaries or to actual adversaries who are not present to defend themselves, counterattack, etc. What we have here is the specialized concept ONE-PARTY RATIONAL ARGUMENT.

Finally, there is a distinction between an argument as a *process* (arguing) and an argument as a *product* (what has been written or said in the course of arguing). In this case, the process and the product are intimately related aspects of the same general concept, neither of which can exist without the other, and either of which can be focused on. Thus we speak of the stage of an argument as applying indifferently to the process or the product.

A ONE-PARTY RATIONAL ARGUMENT is a specialized branch of the general concept ARGUMENT and, as such, has many special constraints on it. Since there is no particular adversary present, an idealized adversary must be assumed. If the purpose of victory is to be maintained, it must be victory over an idealized adversary who is not present. The only way to guarantee victory is to be able to overcome all possible adversaries and to win neutral parties over to your side. To do this, you have to anticipate possible objections, defenses, attacks, etc., and deal with them as you construct your argument. Since this is a RATIONAL ARGUMENT, all of these steps must be taken, not just to win, but in the service of the higher purpose of understanding.

The further restrictions placed on one-party rational arguments require us to pay special attention to certain aspects of argument which are not so important (or perhaps not even present) in everyday argument. Among them are:

Content: You have to have enough supporting evidence and say enough of the right things in order to make your point and to overcome any possible objections.

Progress: You have to start with generally agreed upon premises and move in linear fashion toward some conclusion.

Structure: RATIONAL ARGUMENT requires appropriate logical connections among the various parts.

Strength: The ability of the argument to withstand assault depends on the weight of the evidence and the tightness of the logical connections.

Basicness: Some claims are more important to maintain and defend than others, since subsequent claims will be based upon them.

Obviousness: In any argument there will be things which are not obvious. These need to be identified and explored in sufficient detail.

Directness: The force of an argument can depend on how straightforwardly you move from premises to conclusions.

Clarity: What you are claiming and the connections between your claims must be sufficiently clear for the reader to understand them.

These are aspects of a one-party rational argument that are not necessarily present in an ordinary everyday argument. The concept CONVERSATION and the ARGUMENT IS WAR metaphor do not focus on these aspects, which are crucial to idealized RATIONAL ARGUMENT. As a result, the concept RATIONAL ARGUMENT is further defined by means of other metaphors which *do* enable us to focus on these important aspects: AN ARGUMENT IS A JOURNEY, AN ARGUMENT IS A CONTAINER, and AN ARGUMENT IS A BUILDING. As we will see, each of these gives us a handle on some of the above aspects of the concept RATIONAL ARGUMENT. No one of them is sufficient to give us a complete, consistent, and comprehensive understanding of all these aspects, but together they do the job of giving us a coherent understanding of what a rational argument is. We will now take up the question of what it means for various different metaphors, each of which partially structures a concept, to jointly provide a coherent understanding of the concept as a whole.

Coherence within a Single Metaphor

We can get some idea of the mechanism of coherence within a single metaphorical structuring by starting with the metaphor AN ARGUMENT IS A JOURNEY. This metaphor has to do with the goal of the argument, the fact that it must have a beginning, proceed in a linear fashion, and make progress in stages toward that goal. Here are some obvious instances of the metaphor:

AN ARGUMENT IS A JOURNEY

We have *set out* to prove that bats are birds.
When we get to the next point, we shall see that philosophy is dead.
So far, we've seen that no current theories will work.
We will *proceed* in a *step-by-step* fashion.
Our *goal* is to show that hummingbirds are essential to military defense.
This observation *points the way to* an elegant solution.
We have *arrived at* a disturbing conclusion.

One thing we know about journeys is that a JOURNEY DEFINES A PATH.

A JOURNEY DEFINES A PATH

He *strayed from* the path.
He's *gone off in the wrong direction*.
They're *following* us.
I'm *lost*.

Putting together AN ARGUMENT IS A JOURNEY and A JOURNEY DEFINES A PATH, we get:

AN ARGUMENT DEFINES A PATH

He *strayed from the line* of argument
Do you *follow* my argument?
Now we've *gone off in the wrong direction* again.
I'm *lost*.
You're *going around in circles*.

Moreover, paths are conceived of as surfaces (think of a carpet unrolling as you go along, thus creating a path behind you):

THE PATH OF A JOURNEY IS A SURFACE

We *covered* a lot of ground.
He's *on* our trail.

He strayed *off* the trail.
We went back *over* the same trail.

Given that AN ARGUMENT DEFINES A PATH and THE PATH OF A JOURNEY IS A SURFACE, we get:

THE PATH OF AN ARGUMENT IS A SURFACE

We have already *covered* those points.
We have *covered* a lot of *ground* in our argument.
Let's go back *over* the argument again.
You're getting *off* the subject.
You're really *onto* something there.
We're well *on* our *way* to solving this problem.

Here we have a set of cases that fall under the metaphor AN ARGUMENT IS A JOURNEY. What makes them systematic is a pair of metaphorical entailments that are based on two facts about journeys.

The facts about journeys:

A JOURNEY DEFINES A PATH
THE PATH OF A JOURNEY IS A SURFACE

The metaphorical entailments:

AN ARGUMENT IS A JOURNEY
A JOURNEY DEFINES A PATH
Therefore, AN ARGUMENT DEFINES A PATH

AN ARGUMENT IS A JOURNEY
THE PATH OF A JOURNEY IS A SURFACE
Therefore, THE PATH OF AN ARGUMENT IS A SURFACE

Here metaphorical entailments characterize the *internal* systematicity of the metaphor AN ARGUMENT IS A JOURNEY, that is, they make coherent all the examples that fall under that metaphor.

Coherence between Two Aspects of a Single Concept

AN ARGUMENT IS A JOURNEY is only one of the metaphors for arguments, the one we use to highlight or talk about the

goal, direction, or progress of an argument. When we want
to talk about the content of an argument, we use the struc-
turally complex metaphor AN ARGUMENT IS A CONTAINER.
Containers can be viewed as defining a limited space (with a
bounding surface, a center, and a periphery) and as holding
a substance (which may vary in amount, and which may
have a core located in the center). We use the ARGUMENT IS
A CONTAINER metaphor when we want to highlight any of
these aspects of an argument.

AN ARGUMENT IS A CONTAINER
> Your argument doesn't have much *content*.
> That argument *has holes in it*.
> You don't have *much of* an argument, but his objections have
> even *less substance*.
> Your argument is *vacuous*.
> I'm tired of your *empty* arguments.
> You won't *find* that idea *in* his argument.
> That conclusion *falls out of* my argument.
> Your argument *won't hold water*.
> Those points are *central* to the argument—the rest is *periph-
> eral*.
> I still haven't gotten to the *core* of his argument.

Since the purposes of the JOURNEY and CONTAINER
metaphors are different, that is, since they are used to focus
in detail on different aspects of an argument (goal and progress versus content), we would not expect these metaphors
to overlap completely. It is possible in some cases to focus
jointly on both the JOURNEY (progress) and CONTAINER
(content) aspects of an argument. Thus we get certain
mixed metaphors that display both of these aspects at once.

Overlap between JOURNEY and CONTAINER metaphors:
> *At this point* our argument doesn't have *much content*.
> *In* what we've done *so far*, we have provided the *core* of our
> argument.
> If we keep *going the way we're going*, we'll *fit all the facts in*.

What makes this overlap possible is that the JOURNEY and
CONTAINER metaphors have shared entailments. Both

metaphors allow us to distinguish the form of the argument from the content. In the JOURNEY metaphor, the path corresponds to the form of the argument and the ground covered corresponds to the content. When we are going around in circles, we may have a long path, but we don't cover much ground; that is, the argument doesn't have much content. In a good argument, however, each element of form is used to express some content. In the JOURNEY metaphor, the longer the path (the longer the argument), the more ground is covered (the more content the argument has). In the CONTAINER metaphor, the bounding surface of the container corresponds to the form of the argument, and what is in the container corresponds to the "content" of the argument. In a container that is designed and used most efficiently, all of the bounding surface is used to hold content. Ideally, the more surface there is (the longer the argument), the more substance there is in the container (the more content the argument has). As the path of the journey unfolds, more and more of the surface defined by that path is created, just as more and more of the surface of the container is created. The overlap between the two metaphors is the progressive creation of a surface. As the argument covers more ground (via the JOURNEY surface), it gets more content (via the CONTAINER surface).

What characterizes this overlap is a shared entailment that arises in the following way.

A nonmetaphorical entailment about journeys:
As we make a journey, more of a path is created.
A PATH IS A SURFACE.
Therefore, As we make a journey, more of a surface is created.

A metaphorical entailment about arguments (based on journeys):
AN ARGUMENT IS A JOURNEY.
As we make a journey, more of a surface is created.
Therefore, As we make an argument, more of a surface is created.

A metaphorical entailment about arguments (based on containers):

AN ARGUMENT IS A CONTAINER.
As we make a container, more of a surface is created.
Therefore, As we make an argument, more of a surface is created.

Here the two metaphorical entailments have the same conclusion. This can be represented by the accompanying diagram.

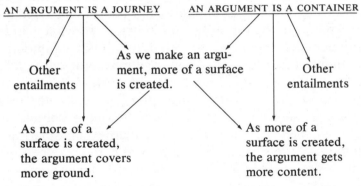

It is this overlap of entailments between the two metaphors that defines the coherence between them and provides the link between the amount of ground the argument covers and the amount of content it has. This is what allows them to "fit together," even though they are not completely consistent, that is, there is no "single image" that completely fits both metaphors. The surface of a container and the surface of the ground are both surfaces by virtue of common topological properties. But our image of ground surface is very different than our images of various kinds of container surfaces. The abstract topological concept of a surface which forms the overlap between these two metaphors is not concrete enough to form an image. In general when metaphors are coherent but not consistent, we should not expect them to form consistent images.

The difference between coherence and consistency is

crucial. Each metaphor focuses on one aspect of the concept ARGUMENT: in this, each serves a single purpose. Moreover, each metaphor allows us to understand one aspect of the concept in terms of a more clearly delineated concept, e.g., JOURNEY or CONTAINER. The reason we need two metaphors is because there is no one metaphor that will do the job—there is no one metaphor that will allow us to get a handle simultaneously on both the direction of the argument and the content of the argument. These two purposes cannot both be served at once by a single metaphor. And where the purposes won't mix, the metaphors won't mix. Thus we get instances of impermissible mixed metaphors resulting from the impossibility of a single clearly delineated metaphor that satisfies both purposes at once. For example, we can speak of the *direction* of the argument and of the *content* of the argument but not of the *direction of the content* of the argument nor of the *content of the direction* of the argument. Thus we do not get sentences like:

> We can now follow the *path* of the *core* of the argument.
> The *content* of the argument *proceeds* as follows.
> The *direction* of his argument has no *substance*.
> I am disturbed by the *vacuous path* of your argument.

The two metaphors would be consistent if there were a way to *completely* satisfy both purposes with one clearly delineated concept. Instead, what we get is coherence, where there is a partial satisfaction of both purposes. For instance, the JOURNEY metaphor highlights both direction and progress toward a goal. The CONTAINER metaphor highlights the content with respect to its amount, density, centrality, and boundaries. The *progress* aspect of the JOURNEY metaphor and the *amount* aspect of the CONTAINER metaphor can be highlighted simultaneously because the amount increases as the argument progresses. And, as we saw, this results in permissible mixed metaphors.

So far we have looked at the coherences between two
metaphorical structurings of the concept ARGUMENT, and
we have found the following:

—Metaphorical entailments play an essential role in linking all of
the instances of a *single* metaphorical structuring of a concept
(as in the various instances of the AN ARGUMENT IS A JOURNEY
metaphor).

—Metaphorical entailments also play an essential role in linking
two different metaphorical structurings of a single concept (as
in the JOURNEY and CONTAINER metaphors for ARGUMENT).

—A shared metaphorical entailment can establish a cross-
metaphorical correspondence. For example, the shared entailment AS WE MAKE AN ARGUMENT, MORE OF A SURFACE IS
CREATED establishes a correspondence between the amount of
ground covered in the argument (which is in the JOURNEY
metaphor) and the amount of content in the argument (which is
in the CONTAINER metaphor).

—The various metaphorical structurings of a concept serve different purposes by highlighting different aspects of the concept.

—Where there is an overlapping of purposes, there is an overlapping of metaphors and hence a coherence between them.
Permissible mixed metaphors fall into this overlap.

—In general, complete consistency across metaphors is rare; coherence, on the other hand, is typical.

17

Complex Coherences across Metaphors

The most important thing to bear in mind throughout our discussion of coherence is the role of purpose. A metaphorical structuring of a concept, say the JOURNEY metaphor for arguments, allows us to get a handle on one aspect of the concept. Thus a metaphor works when it satisfies a purpose, namely, understanding an aspect of the concept. When two metaphors successfully satisfy two purposes, then overlaps in the purposes will correspond to overlaps in the metaphors. Such overlaps, we claim, can be characterized in terms of shared metaphorical entailments and the cross-metaphorical correspondences established by them.

We saw this in a simple example in the last chapter. We would now like to show that the same mechanisms are involved in complex examples. There are two sources of such complexity: (1) there are often many metaphors that partially structure a single concept and (2) when we discuss one concept, we use other concepts that are themselves understood in metaphorical terms, which leads to further overlapping of metaphors. We can isolate the factors that lead to such complexities by examining further the concept ARGUMENT.

In general, arguments serve the purpose of understanding. We construct arguments when we need to show the connections between things that are obvious—that we take for granted—and other things that are not obvious. We do this by putting ideas together. These ideas constitute the content of the argument. The things we take for granted are the starting point of the argument. The things we wish to

show are the goals that we must reach. As we proceed
toward these goals, we make progress by establishing connections. The connections may be strong or weak, and the
network of connections has an overall structure. In any
argument certain ideas and connections may be more basic
than others, certain ideas will be more obvious than others.
How good an argument is will depend on its content, the
strength of the connections, how directly it establishes the
connections, and how easy it is to understand the connections. Briefly, the various ARGUMENT metaphors serve
the purpose of providing an understanding of the following
aspects of the concept:

content	basicness
progress	obviousness
structure	directness
strength	clarity

In the preceding chapter we saw that the JOURNEY
metaphor focuses at least on content and progress, that the
CONTAINER metaphor focuses at least on content, and that
there is an overlap based on the progressive accumulation
of content. But these two metaphors serve even more purposes and are involved in even more complex coherences.
We can see this by considering a third metaphor for arguments:

AN ARGUMENT IS A BUILDING

We've got the *framework* for a *solid* argument.

If you don't *support* your argument with *solid* facts, the
whole thing will *collapse*.

He is trying to *buttress* his argument with a lot of irrelevant
facts, but it is still so *shaky* that it will easily *fall apart*
under criticism.

With the *groundwork* you've got, you can *construct* a pretty
strong argument.

Together, the JOURNEY, CONTAINER, and BUILDING metaphors focus on all of the above aspects of the concept ARGU-
MENT, as the following lists show:

JOURNEY	CONTAINER	BUILDING
content	content	content
progress	progress	progress
directness	basicness	basicness
obviousness	strength	strength
	clarity	structure

Here are some examples of how we understand each of these aspects in terms of the metaphors:

JOURNEY

So far, we haven't *covered much ground.* (*progress, content*)
This is a *roundabout* argument. (*directness*)
We need to *go into this further* in order to *see clearly* what's involved. (*progress, obviousness*)

CONTAINER

You have all the right ideas *in* your argument, but the argument is *still* not *transparent.* (*content, progress, clarity*)
These ideas form the *solid core* of the argument. (*strength, basicness*)

BUILDING

We've got a *foundation* for the argument, now we need a *solid framework.* (*basicness, strength, structure*)
We *have now constructed most of the argument.* (*progress, content*)

We saw in the preceding chapter that the fact that both journeys and containers define surfaces was the basis for the overlap between the JOURNEY and CONTAINER metaphors. The fact that a building also has a surface, namely, the foundation and the outer shell, makes possible further overlaps with the BUILDING metaphor. In each case the *surface* defines the *content,* but in different ways:

JOURNEY: The surface defined by the path of the argument "covers ground," and the content is the ground covered by the argument.

CONTAINER: The content is inside the container, whose boundaries are defined by its surface.

BUILDING: The surface is the outer shell and foundation, which define an interior for the building. But in the BUILDING metaphor, unlike the CONTAINER metaphor, the content is not *in* the interior; instead, the foundation and outer shell *constitute* the content. We can see this in examples like: "The foundation of your argument does not have enough content to support your claims" and "The framework of your argument does not have enough substance to withstand criticism."

Let us call these surfaces "content-defining surfaces."

The notion of a content-defining surface is not sufficient to account for many of the coherences that we find among the metaphors. For example, there are instances of metaphorical overlap based on the notion of depth. Since depth is also defined relative to a surface, we might think that the depth-defining surface for each metaphor would be the same as the content-defining surface. However, this is not always the case, as the following examples show:

This is a *shallow* argument; it needs more *foundation*. (BUILDING)
We have *gone over* these ideas *in great depth*. (JOURNEY)
You haven't gotten to the *deepest* points yet—those at the *core* of the argument. (CONTAINER)

In both the BUILDING and JOURNEY metaphors, the depth-defining surface is the ground level. In the CONTAINER metaphor, it is again the container surface.

	JOURNEY	CONTAINER	BUILDING
Content-defining surface	Surface created by path (the cover)	Surface of the container	Foundation and shell
Depth-defining surface	Ground level	Surface of the container	Ground level

Before proceeding to the coherences, it is important to recognize that there are two different notions of depth operating here. In the BUILDING and CONTAINER metaphors, what is deeper is more basic. The most basic parts of the argument are the deepest: the foundation and the core. However, in the JOURNEY metaphor, deep facts are those that are not obvious. Facts that are not on the surface are hidden from immediate view; we need to go into them in depth. The purposes of an argument include covering certain topics (finishing with them—"putting the lid on") and, in addition, covering them at *appropriate depths*. Progress in an argument is not merely a matter of covering topics; it also requires us to go sufficiently deeply into them. Going into the topic to the required depth is part of the journey:

> As we *go into* the topic *more deeply,* we find...
> We have *come to a point* where we must *explore* the issues at a *deeper level.*

Since most of the journey is over the surface of the earth, it is that surface that defines the *depth* of the topics to be covered. But as we go into any one topic in depth, we leave a trail (a surface) behind us, as we do on all parts of the journey. It is by leaving this surface behind that we *cover* a topic *at a certain depth.* This accounts for the following expressions:

> We will be *going deeply into* a variety of topics.
> *As we go along,* we will go through these issues *in depth.*
> We have now *covered* all the topics *at the appropriate levels.*

Thus the metaphorical orientation of depth corresponds to basicness in the BUILDING and CONTAINER metaphors but to lack of obviousness in the JOURNEY metaphor. Since depth and progress are very different aspects of an argument, there is no *consistent* image possible within any of the ARGUMENT metaphors. But here, as before, though consistency is not possible, there is metaphorical coherence.

Having clarified the distinction between content-defining surfaces and depth-defining surfaces, we are in a position to see a number of other complex coherences. As in the case of the coherence between the JOURNEY and CONTAINER metaphors, there is coherence among all three metaphors based on the fact that all three have content-defining surfaces. As the argument proceeds, more of a surface is created, and hence the argument gets more content. This overlap among the three metaphorical structurings of the concept allows mixed metaphors of the following sort:

> *So far* we have *constructed* the *core* of our argument.

Here "so far" is from the JOURNEY metaphor, "construct" is from the BUILDING metaphor, and "core" is from the CONTAINER metaphor. Notice that we can say pretty much the same thing by using the building concept "foundation" or the neutral concept "most basic part" in place of "core":

> *So far* we have *constructed* the *foundation* of the argument.
> *So far* we have *constructed* the *most basic part* of the argument.

What makes this possible is that depth characterizes basicness in both the BUILDING and CONTAINER metaphors. Both of them have a deepest, that is, most basic part: In the CONTAINER metaphor it is the *core,* and in the BUILDING metaphor it is the *foundation.* Thus we have a correspondence between the two metaphors. This can be seen in the following examples, where the CONTAINER and BUILDING metaphors can be freely mixed by virtue of the correspondence.

> These points are *central* to our argument and provide the *foundation* for all that is to come.
> We can *undermine* the argument by showing that the *central* points in it are weak.
> The most important ideas, *upon* which everything else *rests,* are at the *core* of the argument.

The correspondence here is based on the shared entailment:

> AN ARGUMENT IS A BUILDING.
> A building has a deepest part.
> _____
> Therefore, AN ARGUMENT HAS A DEEPEST PART.

AN ARGUMENT IS A CONTAINER.

A container has a deepest part.

Therefore, AN ARGUMENT HAS A DEEPEST PART.

Since depth characterizes basicness for both metaphors, the deepest part is the most basic part. The concept MOST BASIC PART therefore falls into the overlap of the two metaphors and is neutral between them.

Since the purpose of an argument is to provide understanding, it is not surprising that the metaphor UNDERSTANDING IS SEEING should overlap with the various ARGUMENT metaphors. When you travel, you see more as you go along. This carries over to the metaphor AN ARGUMENT IS A JOURNEY. As you go along through the argument, you see more—and, since UNDERSTANDING IS SEEING, you understand more. This accounts for expressions like:

> We have just *observed* that Aquinas used certain Platonic notions.
> *Having come this far*, we can now *see* how Hegel went wrong.

Because a journey may have a guide who points out things of interest along the way, we also get expressions like:

> We will now *show* that Green misinterpreted Kant's account of will.
> *Notice* that X does not follow from Y without added assumptions.
> We ought to *point out* that no such proof has yet been found.

In these cases, the author is the guide who takes us through the argument.

Part of the JOURNEY metaphor involves going deeply into a subject. The UNDERSTANDING IS SEEING metaphor applies in this case too. In an argument the superficial points (those on the surface) are obvious; they are easy to see, easy to understand. But the deeper points are not obvious. It requires effort—digging—to reveal them so that we can see them. As we go more deeply into an issue, we reveal more, which allows us to see more, that is, to understand more. This accounts for expressions like:

Dig further into his argument and you will *discover* a great deal.
We can *see* this only if we *delve deeply into* the issues.
Shallow arguments are practically worthless, since they don't *show* us very much.

The UNDERSTANDING IS SEEING metaphor also overlaps with the BUILDING metaphor, where what is seen is the structure (shape, form, outline, etc.) of the argument:

We can now *see* the *outline* of the argument.
If we *look* carefully at the *structure* of the argument...

Finally, the UNDERSTANDING IS SEEING metaphor overlaps with the CONTAINER metaphor, where what we see is the content (through the surface of the container), as in:

That is a remarkably *transparent* argument.
I didn't *see* that point *in* your argument.
Since your argument isn't very *clear,* I can't *see* what you're getting at.
Your argument has no *content* at all—I can *see right through* it.

Another cross-metaphorical coherence appears in discussing the quality of an argument. Many of the aspects of an argument that are focused on by the various ARGUMENT metaphors can be quantified—for example, content, clarity, strength, directness, and obviousness. The MORE IS BETTER metaphor overlaps with all of the ARGUMENT metaphors and allows us to view quality in terms of quantity. Thus we have examples like the following:

That's *not much of an argument.*
Your argument *doesn't have any content.*
It's not a very good argument, since it *covers hardly any ground* at all.
This argument won't do—it's just *not clear enough.*
Your argument is *too weak* to support your claims.

The argument is *too roundabout*—no one will be able to follow it.

Your argument doesn't cover the subject matter *in enough depth*.

All of these assess quality in terms of quantity.

We have by no means exhausted all the cross-metaphorical coherences involving ARGUMENT metaphors. Consider, for example, the extensive network of coherences based on the ARGUMENT IS WAR metaphor. Here it is possible to win or lose, to attack and defend, to plan and pursue a strategy, etc. Here arguments may be fortresses via the BUILDING metaphor, so that we can launch an attack on an argument, knock holes in it, tear it down and destroy it. Arguments may also be missiles, via the CONTAINER metaphor. Thus we can offer the challenge "Shoot!" and the argument in reply may be right on target and hit the mark. In defense you can try to shoot down your opponent's argument.

By now it should be clear that the same kinds of coherence found in simple examples also occur in far more complex cases of the sort we have just examined. What may at first appear to be random, isolated metaphorical expressions—for example, *cover those points, buttress your argument, get to the core, dig deeper, attack a position,* and *shoot down*—turn out to be not random at all. Rather, they are part of whole metaphorical systems that together serve the complex purpose of characterizing the concept of an argument in all of its aspects, as we conceive them. Though such metaphors do not provide us with a single consistent concrete image, they are nonetheless coherent and do fit together when there are overlapping entailments, though not otherwise. The metaphors come out of our clearly delineated and concrete experiences and allow us to construct highly abstract and elaborate concepts, like that of an argument.

18

Some Consequences for Theories of Conceptual Structure

Any adequate theory of the human conceptual system will have to give an account of how concepts are (1) grounded, (2) structured, (3) related to each other, and (4) defined. So far we have given a provisional account of grounding, structuring, and relations among concepts (subcategorization, metaphorical entailment, part, participant, etc.) for what we take to be typical cases. We have argued, moreover, that most of our conceptual system is metaphorically structured and have given a brief account of what that means. Before we explore the implications of our views for definition, we need to look at two major strategies that linguists and logicians have used to handle, without any reference to metaphor, what we have called metaphorical concepts.

The two strategies are *abstraction* and *homonymy*. To see how these differ from the account we have offered, consider the word *buttress* in "He buttressed the wall" and "He buttressed his argument with more facts." On our account, we understand *buttress* in "He buttressed his argument" in terms of the concept BUTTRESS, which is part of the BUILDING gestalt. Since the concept ARGUMENT is comprehended partly in terms of the metaphor AN ARGUMENT IS A BUILDING, the meaning of "buttress" in the concept ARGUMENT will follow from the meaning it has in the concept BUILDING, plus the way that the BUILDING metaphor in general structures the concept ARGUMENT. Thus we do not need an independent definition for the concept BUTTRESS in "He buttressed his argument."

Against this, the *abstraction* view claims that there is a single, very general, and abstract concept BUTTRESS, which is neutral between the BUILDING "buttress" and the ARGUMENT "buttress." According to this view, "He buttressed the wall" and "He buttressed his argument" are both special cases of the same very abstract concept. The *homonymy* view takes the opposite tack. Instead of claiming that there is one abstract and neutral concept BUTTRESS, the homonymy view claims that there are two different and independent concepts, BUTTRESS1 *and* BUTTRESS2. There is a *strong* homonymy view, according to which BUTTRESS1 and BUTTRESS2 are entirely different and have nothing to do with each other, since one refers to physical objects (building parts) and the other to an abstract concept (a part of an argument). The *weak* homonymy view maintains that there are distinct and independent concepts BUTTRESS1 and BUTTRESS2 but allows that their meanings may be similar in some respects and that the concepts are related by virtue of this similarity. It denies, however, that either concept is understood in terms of the other. All it claims is that the two concepts have something in common: an abstract similarity. On this point, the weak homonymy view shares an element with the abstraction view, since the abstract similarity would have precisely the properties of the core concept that is hypothesized by the abstraction theory.

We would now like to show why neither the abstraction nor the homonymy theory can account for the kinds of facts that have led us to the theory of metaphorical concepts—in particular, the facts concerning the metaphorical types (orientational, physical, and structural) and their properties (internal systematicity, external systematicity, grounding, and coherence).

Inadequacies of the Abstraction View

The abstraction theory is inadequate in several respects. First, it does not seem to make any sense at all with respect to UP-DOWN orientation metaphors, such as HAPPY IS UP, CONTROL IS UP, MORE IS UP, VIRTUE IS UP, THE FUTURE IS UP, REASON IS UP, etc. What single general concept with any content at all could be an abstraction of HEIGHT, HAPPINESS, CONTROL, MORE, VIRTUE, THE FUTURE, REASON, and NORTH and would precisely fit them all? Moreover, it would seem that UP and DOWN could not be at the same level of abstraction, since UP applies to the FUTURE, while DOWN does not apply to the PAST. We account for this by partial metaphorical structuring, but under the abstraction proposal UP would have to be more abstract in some sense than DOWN, and that does not seem to make sense.

Second, the abstraction theory would not distinguish between metaphors of the form *A is B* and those of the form *B is A,* since it would claim that there are neutral terms covering both domains. For example, English has the LOVE IS A JOURNEY metaphor but no JOURNEYS ARE LOVE metaphor. The abstraction view would deny that love is understood in terms of journeys, and it would be left with the counterintuitive claim that love and journeys are understood in terms of some abstract concept neutral between them.

Third, different metaphors can structure different aspects of a single concept; for example, LOVE IS A JOURNEY, LOVE IS WAR, LOVE IS A PHYSICAL FORCE, LOVE IS MADNESS. Each of these provides one perspective on the concept LOVE and structures one of many aspects of the concept. The abstraction hypothesis would seek a single general concept LOVE abstract enough to fit all of these aspects. Even if this were possible, it would miss the point that these metaphors are not jointly characterizing a core concept LOVE but are separately characterizing different aspects of LOVE.

Fourth, if we look at structural metaphors of the form *A is B* (e.g., LOVE IS A JOURNEY, THE MIND IS A MACHINE, IDEAS ARE FOOD, AN ARGUMENT IS A BUILDING), we find that *B* (the defining concept) is more clearly delineated in our experience and typically more concrete than *A* (the defined concept). Moreover, there is always more in the defining concept than is carried over to the defined concept. Take IDEAS ARE FOOD. We may have *raw facts* and *half-baked ideas*, but there are no *sautéed, broiled,* or *poached ideas*. In AN ARGUMENT IS A BUILDING only the foundation and outer shell play a part in the metaphor, not the inner rooms, corridors, roof, etc. We have explained this asymmetry in the following way: the less clearly delineated (and usually less concrete) concepts are partially understood in terms of the more clearly delineated (and usually more concrete) concepts, which are directly grounded in our experience. The abstraction view has no explanation for this asymmetry, since it cannot explain the tendency to understand the less concrete in terms of the more concrete.

Fifth, under the abstraction proposal there are no metaphorical concepts at all and, therefore, no reason to expect the kind of systematicity that we have found. Thus, for example, there is no reason to expect a whole system of food concepts to apply to ideas or a whole system of building concepts to apply to arguments. There is no reason to expect the kind of internal consistency that we found in the TIME IS A MOVING OBJECT cases. In general, the abstraction view cannot explain the facts of internal systematicity.

Abstraction also fails to explain external systematicity. Our proposal accounts for the way that various metaphors for a single concept (e.g., the JOURNEY, BUILDING, CONTAINER, and WAR metaphors for arguments) overlap in the way that they do. This is based on the shared purposes and shared entailments of the metaphorical concepts. The way that individual concepts (such as CORE, FOUNDATION, COVER, SHOOT DOWN, etc.) mix with each other is predicted on the basis of shared purposes and entailments in the entire metaphorical system. Since the abstraction proposal does not have any metaphorical systems, it cannot explain why metaphors can mix the way they do.

Sixth, since the abstraction proposal has no partial

metaphorical structuring, it cannot account for metaphorical extensions into the unused part of the metaphor, as in "Your theory is constructed out of cheap stucco" and many others that fall within the unused portion of the THEORIES ARE BUILDINGS metaphor.

Finally, the abstraction hypothesis assumes, in the case of LOVE IS A JOURNEY, for example, that there is a set of abstract concepts, neutral with respect to love and journeys, that can "fit" or "apply to" both of them. But in order for such abstract concepts to "fit" or "apply to" love, the concept LOVE must be independently structured so that there can be such a "fit." As we will show, LOVE is not a concept that has a clearly delineated structure; whatever structure it has it gets only via metaphors. But the abstraction view, which has no metaphors to do the structuring, must assume that a structure as clearly delineated as the relevant aspects of journeys exists independently for the concept LOVE. It's hard to imagine how.

Inadequacies of the Homonymy View

Strong Homonymy

Homonymy is the use of the same word for different concepts, as in the *bank* of a river and the *bank* you put your money in. Under the *strong* homonymy theory of the kinds of examples we have been considering, the word "attack" in "They *attacked* the fort" and "They *attacked* my argument" would stand for two entirely different and unrelated concepts. The fact that the same word, "attack," is used would be considered an accident. Similarly, the word "in" of "*in* the kitchen," "*in* the Elks," and "*in* love" would stand for three entirely different, independent, and unrelated concepts—and again it would be accidental that the same word was used. According to this view, English has dozens of separate and unrelated concepts, all accidentally expressed by the word "in." In general, the strong homonymy view cannot account for the relationships that we have identified in systems of metaphorical concepts; that is, it views as accidental all the phenomena that we explain in systematic terms.

In the first place, the strong homonymy position cannot account for any of the internal systematicity that we have described. For example, it would be possible, according to this view, for "I'm feeling up" to mean "I'm happy" and, simultaneously, for "my spirits rose" to mean "I got sadder." Nor can this position account for why the whole system of words used for war should apply in a systematic way to arguments or why a system of food terminology should apply in a systematic way to ideas.

Second, the strong homonymy view has the same problems with cases of external systematicity. That is, it cannot account for the overlap of metaphors and the possibility of mixing. It cannot explain, for example, why the "ground covered" in an argument can refer to the same thing as the "content" of the argument. This holds in general for all the examples of mixing that we have given.

Third, the strong homonymy view cannot explain extensions of the used (or unused) portion of a metaphor, as in "His theories are Gothic and covered with gargoyles." Since that theory has no general metaphors like AN ARGUMENT IS A BUILDING, it must view such cases as random.

Weak Homonymy

The obvious general inadequacy of the strong homonymy view is that it cannot account for any of the systematic relationships that we have found in metaphorical concepts because it sees each concept as not only independent but unrelated to other concepts expressed by the same word. The *weak* homonymy view is superior to the strong view precisely because it does allow for the possibility of such relationships. In particular, it holds that the various concepts expressed by a single word can in many cases be related by similarity. The weak homonymy view takes such similarities as given and assumes that they are sufficient to account for all the phenomena that we have observed, though without the use of any metaphorical structuring.

The most obvious difference between the weak homonymy position and ours is that it has no notion of understanding one thing in terms of another and hence no general metaphorical structuring. The reason for this is that most of those who hold this position are not concerned with how our conceptual system is grounded in experience and how understanding emerges from such grounding. Most of the inadequacies we find in the weak homonymy position have to do with its lack of concern for issues of understanding and grounding. These same inadequacies will, of course, apply also to the strong version of the homonymy position.

First, we have suggested that there is *directionality* in metaphor, that is, that we understand one concept in terms of another. Specifically, we tend to structure the less concrete and inherently vaguer concepts (like those for the emotions) in terms of more concrete concepts, which are more clearly delineated in our experience.

The weak homonymy position would deny that we understand the abstract in terms of the concrete or that we understand concepts of one *kind* in terms of concepts of another *kind* at all. It claims only that we can perceive similarities between various concepts and that such similarities will account for the use of the same words for the concepts. It would deny, for example, that the concept BUTTRESS, when part of the concept ARGUMENT, is understood in terms of the physical concept BUTTRESS as used in BUILDING. It would simply claim that these are two distinct concepts, neither of which is used to understand the other but which happen to have an abstract similarity. Similarly, it would say that all of the concepts for *in* or *up* are not ways of understanding concepts partly in terms of spatial orientation but, rather, are independent concepts related by similarity. On this view, it would be an accident that most of the pairs of concepts that exhibit "similarities" happen to consist of one relatively concrete concept and one relatively abstract concept (as is the case with BUTTRESS). In our account the concrete concept is being used to understand the more abstract concept; in theirs, there would be no reason for there to be more similarities between an abstract and a concrete concept than between two abstract concepts or two concrete concepts.

Second, the claim that such similarities exist is highly questionable. For example, what possible similarities could there be that are shared by all of the concepts that are oriented UP? What similarity could there be between UP, on the one hand, and HAPPINESS, HEALTH, CONTROL, CON-SCIOUSNESS, VIRTUE, RATIONALITY, MORE, etc., on the other? What similarities (which are not themselves metaphorical) could there be between a MIND and a BRIT-TLE OBJECT, or between IDEAS and FOOD? What is there that is not metaphorical about an instant of time in itself that gives it the front-back orientation that we saw in our discussion of the TIME IS A MOVING OBJECT metaphor? On the weak homonymy view, this front-back orientation must be assumed as an inherent property of instants of time if expressions like "follow," "precede," "meet the future head on," "face the future," etc., are to be explained on the basis of inherent conceptual similarity. So far as we can see, there is no reasonable theory of inherent similarity that can account for any of these cases.

Third, we have given an account of metaphorical grounding in terms of systematic correspondences in our experience, for example, being dominant in a fight and being physically up. But there is a difference between correspondences in our experience and similarities, since the correspondence need not be based on any similarity. On the basis of such correspondences in our experience, we can give an account of the range of possible metaphors. The weak homonymy position has no predictive power at all and seeks none. It simply tries to provide an after-the-fact account of what similarities there are. Thus, in the cases where similarities can be found, the weak homonymy position still gives no account of why just those similarities should be there.

To our knowledge, no one *explicitly* holds the strong homonymy position, according to which concepts expressed by the same word (like the two senses of "buttress" or the many senses of "in"), are independent and have no significant relationships. Those who hold the homonymy position tend to identify themselves as holding the weak position, where the interdependencies and interrelationships that are observed between concepts are to be accounted for by similarities based on the inherent nature of the concept. However, to our knowledge, no one has ever begun to provide a detailed account of a theory of similarity that could deal with the wide range of examples we have discussed. Although virtually all homonymy theorists espouse the weak version, in practice there seem to be only strong homonymy theories, since no one has attempted to provide the detailed account of similarity necessary to maintain the weak version of the theory. And there is a good reason why no attempt has been made to give such a detailed account of the kinds of examples we have been discussing. The reason is that such an account would require one to address the issue of how we comprehend and understand areas of experience that are not well-defined in their own terms and must be grasped in terms of other areas of experience. In general, philosophers and linguists have not been concerned with such questions.

19

Definition and Understanding

We have seen that metaphor pervades our normal conceptual system. Because so many of the concepts that are important to us are either abstract or not clearly delineated in our experience (the emotions, ideas, time, etc.), we need to get a grasp on them by means of other concepts that we understand in clearer terms (spatial orientations, objects, etc.). This need leads to metaphorical definition in our conceptual system. We have tried with examples to give some indication of just how extensive a role metaphor plays in the way we function, the way we conceptualize our experience, and the way we speak.

Most of our evidence has come from language—from the meanings of words and phrases and from the way humans make sense of their experiences. Yet students of meaning and dictionary makers have not found it important to try to give a general account of how people understand normal concepts in terms of systematic metaphors like LOVE IS A JOURNEY, ARGUMENT IS WAR, TIME IS MONEY, etc. For example, if you look in a dictionary under "love," you find entries that mention affection, fondness, devotion, infatuation, and even sexual desire, but there is no mention of the way in which we comprehend love by means of metaphors like LOVE IS A JOURNEY, LOVE IS MADNESS, LOVE IS WAR, etc. If we take expressions like "Look how far we've come" or "Where are we now?" there would be no way to tell from a standard dictionary or any other standard account of meaning that these expressions are normal ways of talking about the experience of love in our culture. Hints of the existence of such general metaphors may be given in the secondary or tertiary senses of *other* words. For instance, a hint of the LOVE IS MADNESS metaphor may show up in a tertiary sense of the word "crazy" (= "immoderately fond, infatuated"), but this hint shows up as part of the definition of "crazy" rather than as part of the definition of "love."

What this suggests to us is that dictionary makers and other students of meaning have different concerns than we do. We are concerned primarily with how people understand their experiences. We view language as providing data that can lead to general principles of understanding. The general principles involve whole systems of concepts rather than individual words or individual concepts. We have found that such principles are often metaphoric in nature and involve understanding one kind of experience in terms of another kind of experience.

Bearing this in mind, we can see the main difference between our enterprise and that of dictionary makers and other students of meaning. It would be very strange in a dictionary to see "madness" or "journeying" as senses of "love." They are not senses of "love," any more than "food" is one of the senses of "idea." Definitions for a concept are seen as characterizing the things that are inherent in the concept itself. We, on the other hand, are concerned with how human beings get a handle on the concept—how they understand it and function in terms of it. Madness and journeys give us handles on the concept of love, and food gives us a handle on the concept of an idea.

Such a concern for how we comprehend experience requires a very different concept of definition from the standard one. The principal issue for such an account of definition is what gets defined and what does the defining. That is the issue we turn to next.

The Objects of Metaphorical Definition: Natural Kinds of Experience

We have found that metaphors allow us to understand one domain of experience in terms of another. This suggests that understanding takes place in terms of entire domains of experience and not in terms of isolated concepts. The fact that we have been led to hypothesize metaphors like LOVE IS A JOURNEY, TIME IS MONEY, and ARGUMENT IS WAR suggests to us that the focus of definition is at the level of basic domains of experience like love, time, and argument. These experiences are then conceptualized and defined in terms of other basic domains of experience like journeys, money, and war. The definition of subconcepts, like BUDGETING TIME and ATTACKING A CLAIM, should fall out as consequences of defining the more general concepts (TIME, ARGUMENT, etc.) in metaphorical terms.

This raises a fundamental question: What constitutes a "basic domain of experience"? Each such domain is a structured whole within our experience that is conceptualized as what we have called an *experiential gestalt*. Such gestalts are *experientially basic* because they characterize structured wholes within recurrent human experiences. They represent coherent organizations of our experiences in terms of natural dimensions (parts, stages, causes, etc.). Domains of experience that are organized as gestalts in terms of such natural dimensions seem to us to be *natural kinds of experience*.

They are *natural* in the following sense: These kinds of experiences are a product of

Our bodies (perceptual and motor apparatus, mental capacities, emotional makeup, etc.)

Our interactions with our physical environment (moving, manipulating objects, eating, etc.)

Our interactions with other people within our culture (in terms of social, political, economic, and religious institutions)

In other words, these "natural" kinds of experience *are products of human nature*. Some may be universal, while others will vary from culture to culture.

We are proposing that the concepts that occur in metaphorical definitions are those that correspond to natural kinds of experience. Judging by the concepts that are *defined by* the metaphors we have uncovered so far, the following would be examples of concepts for natural kinds of experience in our culture: LOVE, TIME, IDEAS, UNDERSTANDING, ARGUMENTS, LABOR, HAPPINESS, HEALTH, CONTROL, STATUS, MORALITY, etc. These are concepts that require metaphorical definition, since they are not clearly enough delineated in their own terms to satisfy the purposes of our day-to-day functioning.

Similarly, we would suggest that concepts that are used in metaphorical definitions *to define* other concepts also correspond to natural kinds of experience. Examples are PHYSICAL ORIENTATIONS, OBJECTS, SUBSTANCES, SEEING, JOURNEYS, WAR, MADNESS, FOOD, BUILDINGS, etc. These concepts for natural kinds of experience and objects are structured clearly enough and with enough of the right kind of internal structure to do the job of defining other concepts. That is, they provide the right kind of structure to allow us to get a handle on those natural kinds of experience that are less concrete or less clearly delineated in their own terms.

It follows from this that some natural kinds of experience are partly metaphorical in nature, since metaphor plays an essential role in characterizing the structure of the experience. Argument is an obvious example, since experiencing certain activities of talking and listening as an argument partly requires the structure given to the concept ARGUMENT by the ARGUMENT IS WAR metaphor. The experience of time is a natural kind of experience that is understood almost entirely in metaphorical terms (via the spatialization of TIME and the TIME IS A MOVING OBJECT and TIME IS MONEY metaphors). Similarly, all of the concepts (e.g.,

'CONTROL, STATUS, HAPPINESS) that are oriented by UP-DOWN and other spatialization concepts are grounded in natural kinds of experience that are partly understood in metaphorical terms.

Interactional Properties

We have seen that our conceptual system is grounded in our experiences in the world. Both directly emergent concepts (like UP-DOWN, OBJECT, and DIRECT MANIPULATION) and metaphors (like HAPPY IS UP, EVENTS ARE OBJECTS, ARGU-MENT IS WAR) are grounded in our constant interaction with our physical and cultural environments. Likewise, the dimensions in terms of which we structure our experience (e.g., parts, stages, purposes) emerge naturally from our activity in the world. The kind of conceptual system we have is a product of the kind of beings we are and the way we interact with our physical and cultural environments.

Our concern with the way we understand our experience has led us to a view of *definition* that is very different from the standard view. The standard view seeks to be "objective," and it assumes that experiences and objects have inherent properties and that human beings understand them solely in terms of these properties. Definition for the objectivist is a matter of saying what those inherent properties are by giving necessary and sufficient conditions for the application of the concept. "Love," on the objectivist view, has various senses, each of which can be defined in terms of such inherent properties as fondness, affection, sexual desire, etc. Against this view, we would claim that we comprehend love only partly in terms of such inherent properties. For the most part, our comprehension of love is metaphorical, and we understand it primarily in terms of concepts for other natural kinds of experience: JOURNEYS, MADNESS, WAR, HEALTH, etc. Because defining concepts (JOURNEYS, MADNESS, WAR, HEALTH) emerge from our interactions with one another and with the world, the concept they metaphorically define (e.g., LOVE) will be understood in terms of what we will call *interactional properties*.

In order to get a clearer idea of what interactional properties are in general, let us look at the interactional properties of an object. Take the concept GUN. You might think that such a concept could be characterized entirely in terms of inherent properties of the object itself, for example, its shape, its weight, how its parts are put together, etc. But our concept GUN goes beyond this in ways that can be seen when we apply various modifiers to the concept. For example, take the difference between the modifiers BLACK and FAKE as applied to GUN. The principal difference for objectivist accounts of definition is that a BLACK GUN is a GUN, while a FAKE GUN is *not* a GUN. BLACK is seen as adding an additional property to GUN, while FAKE is seen as applying to the concept GUN to yield another concept that is not a subcategory of GUN. This is about all that is said on the objectivist view. It will allow the entailments:

$$\frac{\text{This is a black gun.}}{\text{Therefore, this is a gun.}} \quad \text{and} \quad \frac{\text{This is a fake gun.}}{\text{Therefore, this is not a gun.}}$$

What such an account does not do is to say what a fake gun *is*. It does not account for entailments like:

$$\frac{\text{This is a fake gun.}}{\text{Therefore, this is not a giraffe.}}$$

$$\frac{\text{This is a fake gun.}}{\text{Therefore, this is not a bowl of bean-sauce noodles.}}$$

And on and on . . .

To account for such an indefinitely long list of entailments, we need a detailed account of just how FAKE modifies the concept GUN. A fake gun has to look enough like a gun for the purpose at hand. That is, it has to have the contextually appropriate perceptual properties of a gun. You have to be able to perform enough of the appropriate physical manipulations that you would with a real gun (e.g., hold it a certain way). In other words, a fake gun has to maintain what we might call *motor-activity properties* of a gun. Moreover, the point of having a fake gun is that it will serve certain of the purposes that a real gun could serve (threatening, being on display, etc.). What makes a fake gun fake is that it cannot *function* like a gun. If it can shoot you, it is a real gun, not a fake gun. Finally, it cannot originally have been made to function like a gun: a broken or inoperable gun is not a fake gun.

Thus, the modifier FAKE preserves certain *kinds* of the properties of GUNS and negates others. To summarize:

FAKE preserves: Perceptual properties (a fake gun looks like a gun)
Motor-activity properties (you handle it like a gun)
Purposive properties (it serves some of the purposes of a gun)

FAKE negates: Functional properties (a fake gun doesn't shoot)
History of function (if it was made to be a real gun, then it's not a fake)

This account of how FAKE affects the concept of GUN indicates that the concept GUN has at least five dimensions, three of which are preserved by FAKE and two of which are negated. This suggests that we conceptualize a gun in terms of a multidimensional gestalt of properties where the dimensions are PERCEPTUAL, MOTOR ACTIVITY, PURPOSIVE, FUNCTIONAL, etc.

If we look at what perceptual, motor-activity, and purposive properties are, we see that they are not inherent in guns themselves. Instead, they have to do with the way we interact with guns. This indicates that the concept GUN, as people actually understand it, is at least partly defined by interactional properties having to do with perception, motor activity, purpose, function, etc. Thus we find that our concepts of objects, like our concepts of events and activities, are characterizable as multidimensional gestalts whose dimensions emerge naturally from our experience in the world.

Categorization

On the standard objectivist view, we can understand (and hence define) an object entirely in terms of a *set* of its *inherent* properties. But, as we have just seen, at least some of the properties that characterize our concept of an object are *interactional*. In addition, the properties do not merely form a *set* but rather a *structured gestalt,* with dimensions that emerge naturally from our experience.

The objectivist account of definition is inadequate to account for understanding in another way as well. On the objectivist view, a category is defined in terms of set theory: it is characterized by a set of inherent properties of the entities in the category. Everything in the universe is either inside or outside the category. The things that are in the category are those that have all the requisite inherent properties. Anything that fails to have one or more of the inherent properties falls outside the category.

This set-theoretical concept of a category does not accord with the way people categorize things and experiences. For human beings, categorization is primarily a means of comprehending the world, and as such it must serve that purpose in a sufficiently flexible way. Set-theoretical categorization, as a model for human categorization, misses the following:

1. As Rosch (1977) has established, we categorize things in terms of prototypes. A prototypical chair, for us, has a well-defined back, seat, four legs, and (optionally) two armrests. But there are nonprototypical chairs as well: beanbag chairs, hanging chairs, swivel chairs, contour chairs, barber chairs, etc. We understand the nonprototypical chairs as being chairs, not just on their own terms, but by virtue of their relation to a prototypical chair.

2. We understand beanbag chairs, barber chairs, and contour chairs as being chairs, not because they share some fixed set of defining properties with the prototype, but rather because they bear a sufficient family resemblance to the prototype. A beanbag chair may resemble a prototypical chair in a different way than a barber chair does. There need be *no fixed core* of properties of prototypical chairs that are shared by both beanbag and barber chairs. Yet they are both chairs because each, in its different way, is sufficiently close to the prototype.

3. Interactional properties are prominent among the kinds of properties that count in determining sufficient family resemblance. Chairs share with stools and other kinds of seats the PURPOSIVE property of allowing us to sit. But the range of MOTOR ACTIVITIES permitted by chairs is usually different from stools and other seats. Thus the interactional properties relevant to our comprehension of chairs will include perceptual properties (the way they look, feel, etc.), functional properties (allowing us to sit), motor-activity properties (what we do with our bodies in getting in and out of them and while we're in them), and purposive properties (relaxing, eating, writing letters, etc.).

4. Categories can be systematically extended in various ways for various purposes. There are modifiers, called hedges (see Lakoff 1975), that pick out the prototype for a category and that define various kinds of relationships to it. Here are a few examples:

> PAR EXCELLENCE: This picks out prototypical members of a category. For example, a robin is a bird par excellence, but chickens, ostriches, and penguins are not birds par excellence.

> STRICTLY SPEAKING: This picks out the nonprototypical cases that ordinarily fall within the category. Strictly speaking, chickens, ostriches, and penguins are birds even though they are not birds par excellence. Sharks, blowfish, catfish, and goldfish are not fish par excellence, but they are fish, strictly speaking.

LOOSELY SPEAKING: This picks out things that are not ordinarily in the category because they lack some central property but which share enough properties so that for certain purposes it could make sense to consider them category members. Strictly speaking, a whale is not a fish, though, loosely speaking, it may be considered one in certain contexts. Strictly speaking, a moped is not a motorcycle, though, loosely speaking, mopeds could be included among motorcycles.

TECHNICALLY: This circumscribes a category relative to some technical purpose. Whether something is technically in the category or not will depend on what the purpose in classifying it is. For the purpose of insurance, a moped is technically not a motorcycle, though for purposes of bridge tolls it technically is.

Some other hedges include *in an important sense, to all intents and purposes, a regular . . . , a veritable . . . , to the extent that . . . , in certain respects,* and many, many more. These various hedges allow us to place objects, events, and experiences under a wide variety of categories for various purposes, e.g., to draw practical distinctions in sensible ways, to provide new perspectives, and to make sense of apparently disparate phenomena.

5. Categories are open-ended. Metaphorical definitions can give us a handle on things and experiences we have already categorized, or they may lead to a recategorization. For example, viewing LOVE as WAR may make sense of certain experiences that you took as LOVE experiences of some kind or other but that you could not fit together in any meaningful way. The LOVE IS WAR metaphor may also lead you to categorize certain experiences as LOVE experiences that you had previously not viewed as such. Hedges also reveal the open-ended nature of our categories; that is, an object may often be seen as being in a category or not, depending on our purposes in classifying it. Though categories are open-ended, categorization is not random, since both metaphors and hedges define (or redefine) categories in systematic ways.

Summary

We have argued that an account of how people understand their experiences requires a view of definition very different from the standard account. An experiential theory of definition has a different notion of what needs to be defined and what does the defining. On our account, individual concepts are not defined in an isolated fashion, but rather in terms of their roles in natural kinds of experiences. Concepts are not defined solely in terms of inherent properties; instead, they are defined primarily in terms of interactional properties. Finally, definition is not a matter of giving some fixed set of necessary and sufficient conditions for the application of a concept (though this may be possible in certain special cases, such as in science or other technical disciplines, though even there it is not always possible); instead, concepts are defined by prototypes and by types of relations to prototypes. Rather than being rigidly defined, concepts arising from our experience are open-ended. Metaphors and hedges are systematic devices for further defining a concept and for changing its range of applicability.

20

How Metaphor Can Give Meaning to Form

We speak in linear order; in a sentence, we say some words earlier and others later. Since speaking is correlated with time and time is metaphorically conceptualized in terms of space, it is natural for us to conceptualize language metaphorically in terms of space. Our writing system reinforces this conceptualization. Writing a sentence down allows us to conceptualize it even more readily as a spatial object with words in a linear order. Thus our spatial concepts naturally apply to linguistic expressions. We know which word occupies the *first position* in the sentence, whether two words are *close* to each other or *far apart,* whether a word is relatively *long* or *short.*

Because we conceptualize linguistic form in spatial terms, it is possible for certain spatial metaphors to apply directly to the *form* of a sentence, as we conceive of it spatially. This can provide automatic direct links between form and content, based on general metaphors in our conceptual system. Such links make the relationship between form and content anything but arbitrary, and some of the meaning of a sentence can be due to the precise form the sentence takes. Thus, as Dwight Bolinger (1977) has claimed, exact paraphrases are usually impossible because the so-called paraphrases are expressed in different forms. We can now offer an explanation for this:

—We spatialize linguistic form.
—Spatial metaphors apply to linguistic form as it is spatialized.
—Linguistic forms are themselves endowed with content by virtue of spatialization metaphors.

More of Form Is More of Content

For example, the CONDUIT metaphor defines a spatial relationship between form and content: LINGUISTIC EXPRESSIONS ARE CONTAINERS, and their meanings are the *content* of those containers. When we see actual containers that are small, we expect their contents to be small. When we see actual containers that are large, we normally expect their contents to be large. Applying this to the CONDUIT metaphor, we get the expectation:

MORE OF FORM IS MORE OF CONTENT.

As we shall see, this is a very general principle that seems to occur naturally throughout the world's languages. Though the CONDUIT metaphor is widespread, we do not know yet whether it is universal. We would expect, however, that some metaphorical spatialization of language would occur in every language and, whatever the details, it would not be surprising to find such correlations of amount.

An English example of MORE OF FORM IS MORE OF CONTENT is iteration:

He ran and ran and ran and ran.

which indicates more running than just

He ran.

Similarly,

He is very very very tall.

indicates that he is taller than

He is very tall.

does. Extended lengthening of a vowel can have the same effect. Saying

He is bi-i-i-i-ig!

indicates that he is bigger than you indicate when you say just

He is big.

Many languages of the world use the morphological device of *reduplication,* that is, the repetition of one or two syllables of a word, or of the whole word, in this way. To our knowledge, all cases of reduplication in the languages of the world are instances where MORE OF FORM stands for MORE OF CONTENT. The most typical devices are:

> Reduplication applied to noun turns singular to plural or collective.
>
> Reduplication applied to verb indicates continuation or completion.
>
> Reduplication applied to adjective indicates intensification or increase.
>
> Reduplication applied to a word for something small indicates diminution.

The generalization is as follows:

> A noun stands for an object of a certain kind.
> More of the noun stands for more objects of that kind.
> A verb stands for an action.
> More of the verb stands for more of the action (perhaps until completion).
> An adjective stands for a property.
> More of the adjective stands for more of the property.
> A word stands for something small.
> More of the word stands for something smaller.

Closeness Is Strength of Effect

A much subtler example of the way metaphor gives meaning to form occurs in English (and possibly in other languages as well, though detailed studies have not been done). English has the conventional metaphor

CLOSENESS IS STRENGTH OF EFFECT.

Thus, the sentence

Who are the men *closest to* Khomeini?

means

Who are the men *who have the strongest effect on* Khomeini?

Here the metaphor has a purely semantic effect. It has to do with the meaning of the word "close." However, *the metaphor can also apply to the syntactic form of a sentence*. The reason is that one of the things the syntax of the sentence indicates is how CLOSE two expressions are to each other. The CLOSENESS is one of *form*.

This metaphor can apply to the relation between form and meaning in the following way:

If the meaning of form *A* affects the meaning of form *B*, then, the CLOSER form *A* is to form *B*, the STRONGER will be the EFFECT of the meaning of *A* on the meaning of *B*.

For example, a sentential negative like *not* has the effect of negating a predicate, as in

John wo*n't* leave until tomorrow.
The form *n't* has the effect of negation on the predicate with the form *leave*.

There is a rule in English, sometimes called *negative transportation*, which has the effect of placing the negative further away from the predicate it logically negates; for example,

Mary does*n't* think he'll *leave* until tomorrow.

Here *n't* logically negates *leave* rather than *think*. This sentence has roughly the same meaning as

Mary thinks he wo*n't* *leave* until tomorrow.

except that in the first sentence, where the negative is FURTHER AWAY from *leave,* it has a WEAKER negative force. In the second sentence, where the negative is CLOSER, the force of negation is STRONGER.

Karl Zimmer (personal communication) has observed that the same principle governs differences like

Harry is not happy

versus

Harry is unhappy.

The negative prefix *un-* is closer to the adjective *happy* than is the separate word *not*. The negative has a stronger effect in *Harry is unhappy* than in *Harry is not happy*. *Unhappy* means *sad,* while *not happy* is open to the interpretation of being neutral—neither happy nor sad, but in between. This is typical of the difference between negatives and negative affixes, both in English and in other languages.

The same metaphor can be seen at work in the following examples:

I *taught* Greek to *Harry*.
I taught *Harry Greek*.

In the second sentence, where *taught* and *Harry* are closer, there is more of a suggestion that Harry actually learned what was taught him—that is, that the teaching had an effect on him. The following examples are even subtler:

I found that the chair was comfortable.
I found the chair comfortable.

The second sentence indicates that I found out that the chair was comfortable *by direct experience*—by sitting in it. The first sentence leaves open the possibility that I found it out *indirectly*—say, by asking people or taking a survey. In the second sentence, the form *I* is CLOSER to the forms *the chair* and *comfortable*. The syntax of the sentence indicates the directness of the experience with the chair by which I found that the chair was comfortable. The CLOSER the form *I* is to the forms *the chair* and *comfortable,* the more direct is the experience that is indicated. Here the effect of the syntax is to indicate the directness of the experience, and CLOSENESS indicates the STRENGTH OF that EFFECT. This

phenomenon in English is verified in detail by Borkin (in press).
   The same metaphor can be seen at work in examples like:

   Sam killed Harry.
   Sam caused Harry to die.

If the cause is a single event, as in the first sentence, the causation is more direct. The second sentence indicates indirect or remote causation—two separate events, Harry's death and what Sam did to cause it. If one wants to indicate causation that is even more indirect, one can say:

   Sam brought it about that Harry died.

The *effect that the syntax has* in these sentences is to indicate *how direct the causal link is* between what Sam did and what happened to Harry. The principle at work is this:

   The CLOSER the form indicating CAUSATION is to the form indicating the EFFECT, the STRONGER the causal link is.

In *Sam killed Harry,* there is only a single form—the word *kill*—to indicate both the CAUSATION and the EFFECT (death). The forms for this meaning are as close as they can be: one word includes them both. This indicates that the causal link is as strong as it could be: a single event. In *Sam caused Harry to die,* there are two separate words—*cause* and *die*—indicating cause and effect. This indicates that the link between the cause and the effect is not as strong as it could be—the cause and the effect are not part of the same event. In *Sam brought it about that Harry died,* there are two separate clauses: *Sam brought it about* and *that Harry died,* which indicates a still weaker causal link.

   In summary, in all of these cases a difference in form indicates a subtle difference in meaning. Just what the subtle differences are is given by the metaphor CLOSENESS IS STRENGTH OF EFFECT, where CLOSENESS applies to elements of the syntax of the sentence, while STRENGTH OF EFFECT applies to the meaning of the sentence. The CLOSE-

NESS has to do with form, while the STRENGTH OF EFFECT has to do with meaning. Thus the metaphor CLOSENESS IS STRENGTH OF EFFECT, which is part of our normal conceptual system, can work either in purely semantic terms, as in the sentence "Who are the men closest to Khomeini?," or it can link *form* to *meaning*, since CLOSENESS can indicate a relation holding between two *forms* in a sentence. The subtle shades of meaning that we see in the examples given above are thus the consequences not of special rules of English but of a metaphor that is in our conceptual system applying naturally to the *form* of the language.

The ME-FIRST Orientation

Cooper and Ross (1975) observe that our culture's view of what a prototypical member of our culture is like determines an orientation of concepts within our conceptual system. The canonical person forms a conceptual reference point, and an enormous number of concepts in our conceptual system are oriented with respect to whether or not they are similar to the properties of the prototypical person. Since people typically function in an *upright* position, see and move *frontward*, spend most of their time performing *actions*, and view themselves as being basically *good*, we have a basis in our experience for viewing ourselves as more UP than DOWN, more FRONT than BACK, more ACTIVE than PASSIVE, more GOOD than BAD. Since we are where we are and exist in the present, we conceive of ourselves as being HERE rather than THERE, and NOW rather than THEN. This determines what Cooper and Ross call the ME-FIRST orientation: UP, FRONT, ACTIVE, GOOD, HERE, and NOW are all oriented toward the canonical person; DOWN, BACKWARD, PASSIVE, BAD, THERE, and THEN are all oriented away from the canonical person.

This cultural orientation correlates with the fact that in

English certain orders of words are more normal than others:

More Normal	Less Normal
up and down	down and up
front and back	back and front
active and passive	passive and active
good and bad	bad and good
here and there	there and here
now and then	then and now

The general principle is: Relative to the properties of the prototypical person, the word whose meaning is NEAREST comes FIRST.

This principle states a correlation between form and content. Like the other principles that we have seen so far, it is a consequence of a metaphor in our normal conceptual system: NEAREST IS FIRST. For example, suppose you are pointing out someone in a picture. If you say

The *first* person on Bill's left is Sam.

you mean

The person who is on Bill's left and *nearest* to him is Sam.

To summarize: Since we speak in linear order, we constantly have to choose which words to put first. Given an otherwise random choice between *up and down* and *down and up,* we automatically choose *up and down.* Of the two concepts UP and DOWN, UP is oriented NEAREST to the prototypical speaker. Since NEAREST IS FIRST is part of our conceptual system, we place the word whose meaning is NEAREST (namely, *up*) in FIRST position. The word order *up and down* is thus more coherent with our conceptual system than the order *down and up.*

For a detailed account of this phenomenon and a discussion of apparent counterexamples, see Cooper and Ross (1975).

Metaphorical Coherence in Grammar

An Instrument Is a Companion

It is common for a child playing with a toy to act toward it as if it were a companion, talking to it, putting it on his pillow next to him at night, etc. Dolls are toys designed especially for this purpose. Behavior like this occurs in adults, who treat certain significant instruments like cars and guns as companions, giving them names, talking to them, etc. Likewise, in our conceptual system, there is the conventional metaphor AN INSTRUMENT IS A COMPANION, which is reflected in the following examples:

AN INSTRUMENT IS A COMPANION

Me and my old Chevy have seen a lot of the country together.

Q: Who's gonna stop me?
A: Me and old Betsy here [said by the cowboy reaching for his gun].

Domenico is going on tour with his priceless Stradivarius.

Sleezo the Magician and his Magic Harmonica will be performing tonight at the Rialto.

Why *With* Indicates Both INSTRUMENTALITY and ACCOMPANIMENT

The word *with* indicates ACCOMPANIMENT in English, as in:

I went to the movies *with* Sally. (COMPANION)

The fact that it is *with* and not some other word that indicates ACCOMPANIMENT is an arbitrary convention of English. In other languages, other words (or grammatical devices like case endings) indicate ACCOMPANIMENT (e.g., *avec* in French). But given the fact that *with* indicates ACCOMPANIMENT in English, it is no accident that *with* also indicates INSTRUMENTALITY, as in:

I sliced the salami *with* a knife. (INSTRUMENT)

The reason that this is not arbitrary is that our conceptual system is structured by the metaphor AN INSTRUMENT IS A COMPANION. It is a *systematic,* not an *accidental,* fact about English that the same word that indicates ACCOMPANI-MENT also indicates INSTRUMENTALITY. This grammatical fact about English is *coherent* with the conceptual system of English.

As it happens, this is not merely a fact about English. With few exceptions, the following principle holds in all the languages of the world:

> The word or grammatical device that indicates ACCOMPANI-MENT also indicates INSTRUMENTALITY.

Since the experiences on which the metaphor AN INSTRU-MENT IS A COMPANION are based are likely to be universal, it is natural that this grammatical principle holds in most languages. Those languages where the principle holds are coherent with the metaphor; those languages where the principle does not hold are not coherent with this metaphor. Where the INSTRUMENT IS A COMPANION coherence does not appear in a language, it is common for some other conceptual coherence to appear in its place. Thus, there are languages where INSTRUMENT is indicated by a form of the verb *use* or where ACCOMPANIMENT is indicated by the word for *and.* These are other, nonmetaphorical, ways in which form may be coherent with content.

The "Logic" of a Language

The use of the same word to indicate INSTRUMENTALITY as well as ACCOMPANIMENT makes sense. It makes such form-content links coherent with the conceptual system of the language. Similarly, the use of spatial words like *in* and *at* for time expressions (e.g., *in* an hour, *at* ten o'clock) makes sense given that TIME is metaphorically conceptualized in terms of SPACE. Metaphors in the conceptual system indicate coherent and systematic relationships between concepts. The use of the same words and grammatical devices for concepts with systematic metaphorical correspondences (like TIME and SPACE) is one of the ways in which the correspondences between form and meaning in a language are "logical" rather than arbitrary.

Conclusion

Subtle Variations in Meaning

Is paraphrase possible? Can two different sentences ever mean exactly the same thing? Dwight Bolinger has spent most of his career showing that this is virtually impossible and that almost any change in a sentence—whether a change in word order, vocabulary, intonation, or grammatical construction—will alter the sentence's meaning, though often in a subtle way. We are now in a position to see *why* this should be so.

We conceptualize sentences metaphorically in spatial terms, with elements of linguistic form bearing spatial properties (like length) and relations (like closeness). Therefore, the spatial metaphors inherent in our conceptual system (like CLOSENESS IS STRENGTH OF EFFECT) will automatically structure relationships between form and content. While some aspects of the meaning of a sentence are consequences of certain relatively arbitrary conventions of the language, other aspects of meaning arise by virtue of our natural attempt to make what we say coherent with our conceptual system. This includes the *form* that we say things in, since that form is conceptualized in spatial terms.

Regularities of Linguistic Form

We have seen that metaphors play an important role in characterizing regularities of linguistic form. One such regularity is the use of the same word to indicate both accompaniment and instrumentality. This regularity is coherent with the conceptual metaphor INSTRUMENTS ARE COMPANIONS. Many of what we perceive as "natural" regularities of linguistic form are regularities that are coherent with metaphors in our conceptual system. Take, for example, the fact that questions typically end in what we perceive as a "rising" intonation, while statements typically end in what we perceive as a "falling" intonation.

This is coherent with the orientational metaphor UN-KNOWN IS UP; KNOWN IS DOWN. This conceptual metaphor can be seen in examples like:

That's still *up in the air*.
I'd like to *raise* some questions about that.
That *settles* the question.
It's still *up* for grabs.
Let's *bring it up* for discussion.

And the reason that the verb *come* is used in *come up with un unswer* is that the answer is conceptualized as starting out DOWN and ending where we are, namely, UP.

Questions typically indicate what is unknown. The use of rising intonation in questions is therefore coherent with UNKNOWN IS UP. The use of falling intonations with statements is therefore coherent with KNOWN IS DOWN. In fact, questions with falling intonation are understood not as real questions but as rhetorical questions indicating statements. For example, "Will you ever learn?" said with falling intonation is a way of saying, indirectly, "You'll never learn." Similarly, statements with rising intonation indicate uncertainty or inability to make sense of something. For example, "Your name's Fred" said with rising intonation indicates that you're not sure and want confirmation. "The Giants traded Madlock" said with rising intonation indicates an inability to make sense of something—that it doesn't fit with what you know. These are all examples of the use of rising and falling intonation coherently with the UNKNOWN IS UP, KNOWN IS DOWN metaphor.

Incidentally, WH-questions in English have falling intonation, for example, "Who did John see yesterday?" Our guess as to the reason for this is that most of the content of

WH-questions is known, and only a single piece of information is taken to be unknown. For instance, "Who did John see yesterday?" presupposes that John saw someone yesterday. As might be expected, tone languages generally do not use intonation to mark questions at all, usually making use of question particles. On the whole, where intonation signals the difference between questions and statements, rising intonation goes with the unknown (yes-no) questions and falling intonation with the known (statements).

Examples like this indicate that regularities of linguistic form cannot be explained in formal terms alone. Many such regularities make sense only when they are seen in terms of the application of conceptual metaphors to our spatial conceptualization of linguistic form. In other words, syntax is not independent of meaning, especially metaphorical aspects of meaning. The "logic" of a language is based on the coherences between the spatialized form of the language and the conceptual system, especially the metaphorical aspects of the conceptual system.

21

New Meaning

The metaphors we have discussed so far are *conventional* metaphors, that is, metaphors that structure the ordinary conceptual system of our culture, which is reflected in our everyday language. We would now like to turn to metaphors that are outside our conventional conceptual system, metaphors that are imaginative and creative. Such metaphors are capable of giving us a new understanding of our experience. Thus, they can give new meaning to our pasts, to our daily activity, and to what we know and believe.

To see how this is possible, let us consider the new metaphor LOVE IS A COLLABORATIVE WORK OF ART. This is a metaphor that we personally find particularly forceful, insightful, and appropriate, given our experiences as members of our generation and our culture. The reason is that it makes our experiences of love coherent—it makes sense of them. We would like to suggest that new metaphors make sense of our experience in the same way conventional metaphors do: they provide coherent structure, highlighting some things and hiding others.

Like conventional metaphors, new metaphors have entailments, which may include other metaphors and literal statements as well. For example, the entailments of LOVE IS A COLLABORATIVE WORK OF ART arise from our beliefs about, and experiences of, what it means for something to be a collaborative work of art. Our personal views of work and art give rise to at least the following entailments for this metaphor:

139

Love is work.
Love is active.
Love requires cooperation.
Love requires dedication.
Love requires compromise.
Love requires a discipline.
Love involves shared responsibility.
Love requires patience.
Love requires shared values and goals.
Love demands sacrifice.
Love regularly brings frustration.
Love requires instinctive communication.
Love is an aesthetic experience.
Love is primarily valued for its own sake.
Love involves creativity.
Love requires a shared aesthetic.
Love cannot be achieved by formula.
Love is unique in each instance.
Love is an expression of who you are.
Love creates a reality.
Love reflects how you see the world.
Love requires the greatest honesty.
Love may be transient or permanent.
Love needs funding.
Love yields a shared aesthetic satisfaction from your joint
   efforts.

Some of these entailments are metaphorical (e.g., "Love is an aesthetic experience"); others are not (e.g., "Love involves shared responsibility"). Each of these entailments may itself have further entailments. The result is a large and coherent network of entailments, which may, on the whole, either fit or not fit our experiences of love. When the network does fit, the experiences form a coherent whole as instances of the metaphor. What we experience with such a metaphor is a kind of reverberation down through the network of entailments that awakens and connects our memories of our past love experiences and serves as a possible guide for future ones.

Let's be more specific about what we mean by "reverberations" in the metaphor LOVE IS A COLLABORATIVE WORK OF ART.

First, the metaphor highlights certain features while suppressing others. For example, the active side of love is brought into the foreground through the notion of WORK both in COLLABORATIVE WORK and in WORK OF ART. This requires the masking of certain aspects of love that are viewed passively. In fact, the emotional aspects of love are almost never viewed as being under the lovers' active control in our conventional conceptual system. Even in the LOVE IS A JOURNEY metaphor, the relationship is viewed as a vehicle that is not in the couple's active control, since it can be *off the tracks,* or *on the rocks,* or *not going anywhere.* In the LOVE IS MADNESS metaphor ("I'm crazy about her," "She's driving me wild"), there is the ultimate lack of control. In the LOVE IS HEALTH metaphor, where the relationship is a patient ("It's a healthy relationship," "It's a sick relationship," "Their relationship is reviving"), the passivity of health in this culture is transferred to love. Thus, in focusing on various aspects of activity (e.g., WORK, CREATION, PURSUING GOALS, BUILDING, HELPING, etc.), the metaphor provides an organization of important love experiences that our conventional conceptual system does not make available.

Second, the metaphor does not merely entail other concepts, like WORK or PURSUING SHARED GOALS, but it entails very specific *aspects* of these concepts. It is not just any work, like working on an automobile assembly line, for instance. It is work that requires that special balance of control and letting-go that is appropriate to artistic creation, since the goal that is pursued is not just any kind of goal but a joint aesthetic goal. And though the metaphor may suppress the out-of-control aspects of the LOVE IS MADNESS metaphor, it highlights another aspect, namely, the sense of almost demonic possession that lies behind our culture's connection between artistic genius and madness.

Third, because the metaphor highlights important love experiences and makes them coherent while it masks other love experiences, the metaphor gives love a new meaning. If those things entailed by the metaphor are for us the most important aspects of our love experiences, then the metaphor can acquire the status of a truth; for many people, love *is* a collaborative work of art. And because it is, the metaphor can have a feedback effect, guiding our future actions in accordance with the metaphor.

Fourth, metaphors can thus be appropriate because they sanction actions, justify inferences, and help us set goals. For example, certain actions, inferences, and goals are dictated by the LOVE IS A COLLABORATIVE WORK OF ART metaphor but not by the LOVE IS MADNESS metaphor. If love is madness, I do not concentrate on what I have to do to maintain it. But if it is work, then it requires activity, and if it is a work of art, it requires a very special *kind* of activity, and if it is collaborative, then it is even further restricted and specified.

Fifth, the meaning a metaphor will have for me will be partly culturally determined and partly tied to my past experiences. The cultural differences can be enormous because each of the concepts in the metaphor under discussion—ART, WORK, COLLABORATION, and LOVE—can vary widely from culture to culture. Thus, LOVE IS A COLLABORATIVE WORK OF ART would mean very different things to a nineteenth-century European Romantic and an Eskimo living in Greenland at the same time. There will also be differences within a culture based on how individuals differ in their views of work and art. LOVE IS A COLLABORATIVE WORK OF ART will mean something very different to two fourteen-year-olds on their first date than to a mature artist couple.

As an example of how the meaning of a metaphor may vary radically within a culture, let us consider some entailments of the metaphor for someone with a view of art very different from our own. Someone who values a work of art not for itself but only as an object for display and someone who thinks that art creates only an illusion, not reality, could see the following as entailments of the metaphor:

Love is an object to be placed on display.
Love exists to be judged and admired by others.
Love creates an illusion.
Love requires hiding the truth.

Because such a person's view of art is different, the metaphor will have a different meaning for him. If his experience of love is pretty much like ours, then the metaphor simply will not fit. In fact, it will be grossly inappropriate. Hence, the same metaphor that gives new meaning to our experiences will not give new meaning to his.

Another example of how a metaphor can create new meaning for us came about by accident. An Iranian student, shortly after his arrival in Berkeley, took a seminar on metaphor from one of us. Among the wondrous things that he found in Berkeley was an expression that he heard over and over and understood as a beautifully sane metaphor. The expression was "the solution of my problems"—which he took to be a large volume of liquid, bubbling and smoking, containing all of your problems, either dissolved or in the form of precipitates, with catalysts constantly dissolving some problems (for the time being) and precipitating out others. He was terribly disillusioned to find that the residents of Berkeley had no such chemical metaphor in mind. And well he might be, for the chemical metaphor is both beautiful and insightful. It gives us a view of problems as things that never disappear utterly and that cannot be solved once and for all. All of your problems are always present, only they may be dissolved and in solution, or they may be in solid form. The best you can hope for is to find a catalyst that will make one problem dissolve without making another one precipitate out. And since you do not have complete control over what goes into the solution, you are constantly finding old and new problems precipitating out and present problems dissolving, partly because of your efforts and partly despite anything you do.

The CHEMICAL metaphor gives us a new view of human problems. It is appropriate to the experience of finding that problems which we once thought were "solved" turn up again and again. The CHEMICAL metaphor says that problems are not the kind of things that can be made to disappear forever. To treat them as things that can be "solved" once and for all is pointless. To live by the CHEMICAL metaphor would be to accept it as a fact that no problem ever disappears forever. Rather than direct your energies toward solving your problems once and for all, you would direct your energies toward finding out what catalysts will dissolve your most pressing problems for the longest time without precipitating out worse ones. The reappearance of a problem is viewed as a natural occurrence rather than a failure on your part to find "the right way to solve it."

To live by the CHEMICAL metaphor would mean that your problems have a different kind of reality for you. A temporary solution would be an accomplishment rather than a failure. Problems would be part of the natural order of things rather than disorders to be "cured." The way you would understand your everyday life and the way you would act in it would be different if you lived by the CHEMICAL metaphor.

We see this as a clear case of the power of metaphor to create a reality rather than simply to give us a way of conceptualizing a preexisting reality. This should not be surprising. As we saw in the case of the ARGUMENT IS WAR metaphor, there are natural kinds of *activity* (e.g., arguing) that are metaphorical in nature. What the CHEMICAL metaphor reveals is that our current way of dealing with problems is another kind of metaphorical activity. At present most of us deal with problems according to what we might call the PUZZLE metaphor, in which problems are PUZZLES for which, typically, there is a correct solution—

and, once solved, they are solved forever. The PROBLEMS ARE PUZZLES metaphor characterizes our present reality. A shift to the CHEMICAL metaphor would characterize a new reality.

But it is by no means an easy matter to change the metaphors we live by. It is one thing to be aware of the possibilities inherent in the CHEMICAL metaphor, but it is a very different and far more difficult thing to live by it. Each of us has, consciously or unconsciously, identified hundreds of problems, and we are constantly at work on solutions for many of them—via the PUZZLE metaphor. So much of our unconscious everyday activity is structured in terms of the PUZZLE metaphor that we could not possibly make a quick or easy change to the CHEMICAL metaphor on the basis of a conscious decision.

Many of our activities (arguing, solving problems, budgeting time, etc.) are metaphorical in nature. The metaphorical concepts that characterize those activities structure our present reality. New metaphors have the power to create a new reality. This can begin to happen when we start to comprehend our experience in terms of a metaphor, and it becomes a deeper reality when we begin to act in terms of it. If a new metaphor enters the conceptual system that we base our actions on, it will alter that conceptual system and the perceptions and actions that the system gives rise to. Much of cultural change arises from the introduction of new metaphorical concepts and the loss of old ones. For example, the Westernization of cultures throughout the world is partly a matter of introducing the TIME IS MONEY metaphor into those cultures.

The idea that metaphors can create realities goes against most traditional views of metaphor. The reason is that metaphor has traditionally been viewed as a matter of mere language rather than primarily as a means of structuring our conceptual system and the kinds of everyday activities we perform. It is reasonable enough to assume that words alone don't change reality. But changes in our conceptual system do change what is real for us and affect how we perceive the world and act upon those perceptions.

The idea that metaphor is just a matter of language and can at best only describe reality stems from the view that what is real is wholly external to, and independent of, how human beings conceptualize the world—as if the study of reality were just the study of the physical world. Such a view of reality—so-called objective reality—leaves out human aspects of reality, in particular the real perceptions, conceptualizations, motivations, and actions that constitute most of what we experience. But the human aspects of reality are most of what matters to us, and these vary from culture to culture, since different cultures have different conceptual systems. Cultures also exist within physical environments, some of them radically different—jungles, deserts, islands, tundra, mountains, cities, etc. In each case there is a physical environment that we interact with, more or less successfully. The conceptual systems of various cultures partly depend on the physical environments they have developed in.

Each culture must provide a more or less successful way of dealing with its environment, both adapting to it and changing it. Moreover, each culture must define a social reality within which people have roles that make sense to them and in terms of which they can function socially. Not surprisingly, the social reality defined by a culture affects its conception of physical reality. What is real for an individual as a member of a culture is a product both of his social reality and of the way in which that shapes his experience of the physical world. Since much of our social reality is understood in metaphorical terms, and since our conception of the physical world is partly metaphorical, metaphor plays a very significant role in determining what is real for us.

22

The Creation of Similarity

We have seen that many of our experiences and activities are metaphorical in nature and that much of our conceptual system is structured by metaphor. Since we see similarities in terms of the categories of our conceptual system and in terms of the natural kinds of experiences we have (both of which may be metaphorical), it follows that many of the similarities that we perceive are a result of conventional metaphors that are part of our conceptual system. We have already seen this in the case of *orientational metaphors*. For example, the orientations MORE IS UP and HAPPY IS UP induce a similarity that we perceive between MORE and HAPPY that we do not see between LESS and HAPPY.

Ontological metaphors also make similarities possible. We saw, for example, that the viewing of TIME and LABOR metaphorically as uniform SUBSTANCES allows us to view them both as being similar to physical resources and hence as similar to each other. Thus the metaphors TIME IS A SUBSTANCE and LABOR IS A SUBSTANCE allow us to conceive of time and labor as similar in our culture, since both can be quantified, assigned a value per unit, seen as serving a purposeful end, and used up progressively. Since these metaphors play a part in defining what is real for us in this culture, the similarity between time and labor is both based on metaphor and real for our culture.

Structural metaphors in our conceptual system also induce similarities. Thus, the IDEAS ARE FOOD metaphor establishes similarities between ideas and food. Both can be digested, swallowed, devoured, and warmed over, and both

can nourish you. These similarities do not exist independently of the metaphor. The concept of swallowing
food is independent of the metaphor, but the concept of
swallowing ideas arises only by virtue of the metaphor. In
fact, the IDEAS ARE FOOD metaphor is based on still more
basic metaphors. For example, it is based partly on the
CONDUIT metaphor, according to which IDEAS ARE OBJECTS
and we can get them from outside ourselves. It also assumes the MIND IS A CONTAINER metaphor, which
establishes a similarity between the mind and the body—
both being CONTAINERS. Together with the CONDUIT
metaphor, we get a complex metaphor in which IDEAS ARE
OBJECTS THAT COME INTO THE MIND, just as pieces of food
are objects that come into the body. It is this metaphorically
created similarity between ideas and food that the IDEAS
ARE FOOD metaphor is partly based on. And, as we have
seen, the similarity itself is a consequence of the CONDUIT
metaphor and the MIND IS A CONTAINER metaphor.

The IDEAS ARE FOOD metaphor fits our experience partly
because of this metaphor-induced similarity. The IDEAS ARE
FOOD metaphor is therefore partly grounded via the MIND IS
A CONTAINER and CONDUIT metaphors. As a consequence
of the IDEAS ARE FOOD metaphor, we get new (metaphorical) similarities between IDEAS and FOOD: both can be
swallowed, digested, and devoured, and both can nourish
you. These food concepts give us a way of understanding
psychological processes that we have no direct and well-
defined way of conceptualizing.

Finally, we can see the creation of similarity in *new
metaphors* as well. For example, the metaphor PROBLEMS
ARE PRECIPITATES IN A CHEMICAL SOLUTION is based on the
conventional metaphor PROBLEMS ARE OBJECTS. In addition, the CHEMICAL metaphor adds PROBLEMS ARE SOLID
OBJECTS, which identifies them as the precipitates in a
chemical solution. The similarities thus induced between
problems as we usually experience them and precipitates in
a chemical solution are: they both have a perceptible form

and thus can be identified, analyzed, and acted upon. These similarities are induced by the PROBLEMS ARE SOLID OBJECTS part of the CHEMICAL metaphor. In addition, when a precipitate is dissolved, it appears to be gone because it does not have a perceptible form and cannot be identified, analyzed, and acted upon. However, it may precipitate out again, i.e., recur in solid form, just as a problem may recur. We perceive this similarity between problems and precipitates as a result of the rest of the CHEMICAL metaphor.

A more subtle example of the similarities created by a *new metaphor* can be seen in LOVE IS A COLLABORATIVE WORK OF ART. This metaphor highlights certain aspects of love experiences, downplays others, and hides still others. In particular, it downplays those experiences that fit the LOVE IS A PHYSICAL FORCE metaphor. By "downplaying," we mean that it is consistent with, but does not focus on, experiences of love that could be reasonably described by "There is a magnetism between us," "We felt sparks," etc. Moreover, it hides those love experiences that fit the LOVE IS WAR metaphor because there is no consistent overlap possible between the two metaphors. The collaborative and cooperative aspects of the LOVE IS A COLLABORATIVE WORK OF ART metaphor are inconsistent with (and therefore hide) the aggressive and dominance-oriented aspects of our love experiences as they might be described by "She is my latest conquest," "He surrendered to her," "She overwhelmed me," etc.

By this means, the LOVE IS A COLLABORATIVE WORK OF ART metaphor puts aside some of our love experiences and picks out others to focus on as if they were our only experiences of love. In doing so it induces a set of similarities between the love experiences it highlights and the real or imagined experiences of collaborating on a work of art. These induced similarities are given in our list of entailments ("Love is work," "Love is an aesthetic experience," etc.).

Within the range of highlighted love experiences, each

experience fits at least one of the similarities given in the list of entailments, and probably no one of them fits all the entailments. For example, a particularly frustrating episode would fit "Love regularly brings frustration" but might not fit "Love is an aesthetic experience" or "Love is primarily valued for its own sake." Each entailment thus states a similarity that holds between certain types of love experiences, on the one hand, and certain types of experiences of collaborative works of art, on the other. No one entailment shows the overall similarity between the *entire range* of highlighted love experiences and the range of experiences involved in producing a collaborative work of art. It is only the whole metaphor, with its entire system of entailments, that shows the similarities between the full range of highlighted love experiences and the full range of experiences of producing a collaborative work of art.

Moreover, there is a similarity induced by the metaphor that goes beyond the mere similarities between the two ranges of experience. The additional similarity is a *structural* similarity. It involves the way we understand how the individual highlighted experiences fit together in a coherent way. The coherence is provided by the structure of what we know about producing a collaborative work of art and is reflected in the way the entailments fit together (e.g., some are entailments of WORK, some are entailments of ART, some are entailments of COLLABORATIVE WORK, etc.). It is only this coherent structure that enables us to understand what the highlighted experiences have to do with each other and how the entailments are related to each other. Thus, by virtue of the metaphor, the range of highlighted love experiences is seen as similar *in structure* to the range of experiences of producing a collaborative work of art.

It is this *structural* similarity between the two ranges of experience that allows you to *find coherence* in the range of highlighted love experiences. Correspondingly, it is by virtue of the metaphor that the highlighted range of experiences is picked out as being coherent. Without the metaphor, this range of experiences does not exist for you as being an identifiable and coherent set of experiences. Conceptualizing LOVE as A COLLABORATIVE WORK OF ART brings them into focus as fitting together into a coherent whole.

Moreover, the metaphor, by virtue of giving coherent structure to a range of our experiences, *creates similarities of a new kind*. For example, we might, independently of the metaphor, see a frustrating love experience as similar to a frustrating experience in producing a work of art jointly with someone, since they are both frustrating. In this sense, the frustrating love experience would also be similar to *any* frustrating experience at all. What the metaphor adds to an understanding of the frustrating love experience is that the *kind* of frustration involved is that involved in producing collaborative artworks. The similarity is similarity with respect to the metaphor.

Thus the precise nature of the similarity between the frustrating love experience and the frustrating art experience is perceived only in understanding the love experience in terms of the art experience. Understanding love experiences in terms of what is involved in producing a collaborative work of art is, by our definition, to comprehend that experience in terms of the metaphorical concept LOVE IS A COLLABORATIVE WORK OF ART.

We can summarize the ways in which metaphors create similarities as follows:

1. Conventional metaphors (orientational, ontological, and structural) are often based on correlations we perceive in our experience. For example, in an industrial culture such as ours there is a correlation between the amount of time a task takes and the amount of labor it takes to accomplish the task. This correlation is part of what allows us to view TIME and LABOR metaphorically as RESOURCES and hence to see a similarity between them. It is important to

remember that correlations are not similarities. Metaphors that are based on correlations in our experience define concepts in terms of which we perceive similarities.

2. Conventional metaphors of the structural variety (e.g., IDEAS ARE FOOD) may be based on similarities that arise out of orientational and ontological metaphors. As we saw, for example, IDEAS ARE FOOD is based on IDEAS ARE OBJECTS (ontological) and THE MIND IS A CONTAINER (ontological and orientational). A structural similarity between IDEAS and FOOD is induced by the metaphor and gives rise to metaphorical similarities (ideas and food can be swallowed, digested, and devoured, can provide nourishment, etc.).

3. New metaphors are mostly structural. They can create similarities in the same way as conventional metaphors that are structural. That is, they can be based on similarities that arise from ontological and orientational metaphors. As we saw, PROBLEMS ARE PRECIPITATES IN A CHEMICAL SOLUTION is based on the physical metaphor PROBLEMS ARE SOLID OBJECTS. This metaphor creates similarities between PROBLEMS and PRECIPITATES, since both can be identified, analyzed, and acted upon. The PROBLEMS ARE PRECIPITATES metaphor creates new similarities, namely, problems can appear to be gone (dissolve into solutions) and later reappear (precipitate out).

4. New metaphors, by virtue of their entailments, pick out a range of experiences by highlighting, downplaying, and hiding. The metaphor then characterizes a similarity between the entire range of highlighted experiences and some other range of experiences. For example, LOVE IS A COLLABORATIVE WORK OF ART picks out a certain range of our love experiences and defines a *structural* similarity between the *entire range* of highlighted experiences and the range of experiences involved in producing collaborative works of art. There may be isolated similarities between love and art experiences that are independent of the metaphor, but the metaphor allows us to find coherence in

these isolated similarities in terms of the overall structural similarities induced by the metaphor.

5. Similarities may be similarities with respect to a metaphor. As we saw, the LOVE IS A COLLABORATIVE WORK OF ART metaphor defines a unique *kind* of similarity. For example, a frustrating love experience may be understood as being similar to a frustrating art experience not merely by virtue of being frustrating but as involving the *kind* of frustration peculiar to jointly producing works of art.

Our view that metaphors can create similarities runs counter to the classical and still most widely held theory of metaphor, namely, the *comparison theory*. The comparison theory says:

1. Metaphors are matters of language and not matters of thought or action. There is no such thing as metaphorical thought or action.

2. A metaphor of the form "*A* is *B*" is a linguistic expression whose meaning is the same as a corresponding linguistic expression of the form "*A* is like *B*, in respects *X, Y, Z* . . . ." "Respects *X, Y, Z, . . .*" characterize what we have called "isolated similarities."

3. A metaphor can therefore only describe preexisting similarities. It cannot create similarities.

Though we have given evidence against much of the comparison theory, we accept what we take to be its basic insight, namely, that metaphors can be based on isolated similarities. We differ with the comparison theory by maintaining that:

1. Metaphor is primarily a matter of thought and action and only derivatively a matter of language.

2.a. Metaphors can be based on similarities, though in many cases these similarities are themselves based on conventional metaphors that are not based on similarities. Similarities based on conventional metaphors are nonetheless *real in our culture*, since conventional metaphors partly define what we find real.

2.b. Though the metaphor may be based partly on isolated similarities, we see the important similarities as those created by the metaphor, as described above.

3. The primary function of metaphor is to provide a partial understanding of one kind of experience in terms of another kind of experience. This may involve preexisting isolated similarities, the creation of new similarities, and more.

It is important to bear in mind that the comparison theory most often goes hand in hand with an objectivist philosophy in which all similarities are objective, that is, they are similarities inherent in the entities themselves. We argue, on the contrary, that the only similarities relevant to metaphor are *similarities as experienced by people*. The difference between *objective similarities* and *experiential similarities* is all-important, and is discussed in detail in chapter 27. Briefly, an objectivist would say that objects have the properties they have independently of anyone who experiences them; the objects are *objectively similar* if they share those properties. To an objectivist it would make no sense to speak of metaphors as *"creating* similarities," since that would require metaphors to be able to change the nature of the external world, bringing into existence objective similarities that did not previously exist.

We agree with objectivists on one major point: that things in the world do play a role in constraining our conceptual system. But they play this role *only through our experience of them*. Our experiences will (1) differ from culture to culture and (2) may depend on our understanding one kind of experience in terms of another, that is, our experiences may be metaphorical in nature. Such experiences determine the categories of our conceptual system. And properties and similarities, we maintain, exist and can be experienced only relative to a conceptual system. Thus, the only kind of similarities relevant to metaphors are *experiential*, not *objective*, similarities.

Our general position is that conceptual metaphors are

grounded in *correlations* within our experience. These experiential correlations may be of two types: *experiential cooccurrence* and *experiential similarity*. An example of experiential cooccurrence would be the MORE IS UP metaphor. MORE IS UP is grounded in the cooccurrence of two types of experiences: adding more of a substance and seeing the level of the substance rise. Here there is no experiential similarity at all. An example of experiential similarity is LIFE IS A GAMBLING GAME, where one experiences actions in life as gambles, and the possible consequences of those actions are perceived as winning or losing. Here the metaphor seems to be grounded in experiential similarity. When such a metaphor is extended, we may experience new similarities between life and gambling games.

23

Metaphor, Truth, and Action

In the preceding chapter we suggested the following:

Metaphors have entailments through which they highlight and make coherent certain aspects of our experience.

A given metaphor may be the only way to highlight and coherently organize exactly those aspects of our experience.

Metaphors may create realities for us, especially social realities. A metaphor may thus be a guide for future action. Such actions will, of course, fit the metaphor. This will, in turn, reinforce the power of the metaphor to make experience coherent. In this sense metaphors can be self-fulfilling prophecies.

For example, faced with the energy crisis, President Carter declared "the moral equivalent of war." The WAR metaphor generated a network of entailments. There was an "enemy," a "threat to national security," which required "setting targets," "reorganizing priorities," "establishing a new chain of command," "plotting new strategy," "gathering intelligence," "marshaling forces," "imposing sanctions," "calling for sacrifices," and on and on. The WAR metaphor highlighted certain realities and hid others. The metaphor was not merely a way of viewing reality; it constituted a license for policy change and political and economic action. The very acceptance of the metaphor provided grounds for certain inferences: there was an external, foreign, hostile enemy (pictured by cartoonists in Arab headdress); energy needed to be given top priorities; the populace would have to make sacrifices; if we didn't meet the threat, we would not survive. It is important to realize that this was not the only metaphor available.

Carter's WAR metaphor took for granted our current concept of what ENERGY is, and focused on how to get enough of it. On the other hand, Amory Lovins (1977) observed that there are two fundamentally different ways, or PATHS, to supply our energy needs. He characterized these metaphorically as HARD and SOFT. The HARD ENERGY PATH uses energy supplies that are inflexible, nonrenewable, needing military defense and geopolitical control, irreversibly destructive of the environment, and requiring high capital investment, high technology, and highly skilled workers. They include fossil fuels (gas and oil), nuclear power plants, and coal gasification. The SOFT ENERGY PATH uses energy supplies that are flexible, renewable, not needing military defense or geopolitical control, not destructive of the environment, and requiring only low capital investment, low technology, and unskilled labor. They include solar, wind, and hydroelectric power, biomass alcohol, fluidized beds for burning coal or other combustible materials, and a great many other possibilities currently available. Lovins' SOFT ENERGY PATH metaphor highlights the technical, economic, and sociopolitical *structure* of the energy system, which leads him to the conclusion that the "hard" energy paths—coal, oil, and nuclear power—lead to political conflict, economic hardship, and harm to the environment. But Jimmy Carter is more powerful than Amory Lovins. As Charlotte Linde (in conversation) has observed, whether in national politics or in everyday interaction, people in power get to impose their metaphors.

New metaphors, like conventional metaphors, can have the power to define reality. They do this through a coherent network of entailments that highlight some features of reality and hide others. The acceptance of the metaphor, which forces us to focus *only* on those aspects of our experience that it highlights, leads us to view the entailments of the metaphor as being *true*. Such "truths" may be true, of course, only relative to the reality defined by the metaphor. Suppose Carter announces that his administration has won a major energy battle. Is this claim true or false? Even to address oneself to the question requires accepting at least the central parts of the metaphor. If you do not accept the existence of an external enemy, if you think there is no external threat, if you recognize no field of battle, no targets, no clearly defined competing forces, then the issue of objective truth or falsity cannot arise. But if you see reality as defined by the metaphor, that is, if you do see the energy crisis as a war, then you can answer the question relative to whether the metaphorical entailments fit reality. If Carter, by means of strategically employed political and economic sanctions, forced the OPEC nations to cut the price of oil in half, then you would say that he would indeed have won a major battle. If, on the other hand, his strategies had produced only a temporary price freeze, you couldn't be so sure and might be skeptical.

Though questions of truth do arise for new metaphors, the more important questions are those of appropriate action. In most cases, what is at issue is not the truth or falsity of a metaphor but the perceptions and inferences that follow from it and the actions that are sanctioned by it. In all aspects of life, not just in politics or in love, we define our reality in terms of metaphors and then proceed to act on the basis of the metaphors. We draw inferences, set goals, make commitments, and execute plans, all on the basis of how we in part structure our experience, consciously and unconsciously, by means of metaphor.

24

Truth

Why Care about a Theory of Truth?

Metaphors, as we have seen, are conceptual in nature. They are among our principal vehicles for understanding. And they play a central role in the construction of social and political reality. Yet they are typically viewed within philosophy as matters of "mere language," and philosophical discussions of metaphor have not centered on their conceptual nature, their contribution to understanding, or their function in cultural reality. Instead, philosophers have tended to look at metaphors as out-of-the-ordinary imaginative or poetic linguistic expressions, and their discussions have centered on whether these linguistic expressions can be *true*. Their concern with truth comes out of a concern with objectivity: *truth* for them means *objective, absolute* truth. The typical philosophical conclusion is that metaphors cannot directly state truths, and, if they can state truths at all, it is only indirectly, via some non-metaphorical "literal" paraphrase.

We do not believe that there is such a thing as *objective* (absolute and unconditional) *truth,* though it has been a long-standing theme in Western culture that there is. We do believe that there are *truths* but think that the idea of truth need not be tied to the objectivist view. We believe that the idea that there is absolute objective truth is not only mistaken but socially and politically dangerous. As we have seen, truth is always relative to a conceptual system that is defined in large part by metaphor. Most of our metaphors have evolved in our culture over a long period, but many are imposed upon us by people in power—political leaders, religious leaders, business leaders, advertisers, the media, etc. In a culture where the myth of objectivism is very much alive and truth is always absolute truth, the people who get to impose their metaphors on the culture get to define what we consider to be true—absolutely and objectively true.

It is for this reason that we see it as important to give an account of truth that is free of the myth of objectivism (according to which truth is always absolute truth). Since we see truth as based on understanding and see metaphor as a principal vehicle of understanding, we think that an account of how metaphors can be true will reveal the way in which truth depends upon understanding.

The Importance of Truth in Our Daily Lives

We base our actions, both physical and social, on what we take to be true. On the whole, truth matters to us because it has survival value and allows us to function in our world. Most of the truths we accumulate—about our bodies, the people we interact with, and our immediate physical and social environments—play a role in daily functioning. They are truths so obvious that it takes a conscious effort to become aware of them: where the front door of the house is, what you can and can't eat, where the nearest gas station is, what stores sell the things you need, what your friends are like, what it would take to insult them, what responsibilities you have. This tiny sample suggests the nature and extent of the vast body of truths that play a role in our daily lives.

The Role of Projection in Truth

In order to acquire such truths and to make use of them, we need an understanding of our world sufficient for our needs. As we have seen, some of this understanding is cast in terms of categories that emerge from our direct experience:

orientational categories, concepts like OBJECT, SUBSTANCE, PURPOSE, CAUSE, etc. We have also seen that when the categories that emerge from direct physical experience do not apply, we sometimes project these categories onto aspects of the physical world that we have less direct experience of. For example, we project a front-back orientation in context onto objects that have no intrinsic fronts or backs. Given a medium-sized rock in our visual field and a ball between us and the rock, say, a foot from it, we would perceive the ball as being *in front of* the rock. The Hausas make a different projection than we do and would understand the ball as being *in back of* the rock. Thus, a front-back orientation is not an inherent property of objects like rocks but rather an orientation that we project onto them, and the way we do this varies from culture to culture. Relative to our purposes, we can conceive of things in the world as being containers or not. We can, for example, conceive of a clearing in a forest as being a CONTAINER and understand ourselves as being IN the clearing, or OUT OF it. Being a container is not an inherent property of that place in the woods where the trees are less dense; it is a property that we project onto it relative to the way we function with respect to it. Relative to other perceptions and purposes, we can view the rest of the forest outside the clearing as a different container and perceive ourselves as being IN the forest. And we can do both simultaneously and speak of EMERGING FROM the forest INTO the clearing.

Similarly, our on-off orientation emerges from our direct experience with the ground, floors, and other horizontal surfaces. Typically, we are *on* the ground, floor, etc., if we are standing on it with our bodies erect. We also project on-off orientations onto walls and conceive of a fly as standing *on* the wall if its legs are in contact with it and its head is oriented away from the wall. The same carries over to the fly on the ceiling: we conceive of it as being *on* rather than *under* the ceiling.

As we have also seen, we perceive various things in the

natural world as entities, often projecting boundaries and surfaces on them where no clear-cut boundaries or surfaces exist naturally. Thus we can conceive of a fogbank as an entity that can be *over* the bay (which we conceive as an entity) and *in front of* the mountain (conceived as an entity with a FRONT-BACK orientation). By virtue of these projections, a sentence like "The fog is in front of the mountain" may be *true*. As is typically the case in our daily lives, truth is relative to understanding, and the truth of such a sentence is relative to the normal way we understand the world by projecting orientation and entity structure onto it.

The Role of Categorization in Truth

In order to understand the world and function in it, we have to categorize, in ways that make sense to us, the things and experiences that we encounter. Some of our categories emerge directly from our experience, given the way our bodies are and the nature of our interactions with other people and with our physical and social environments. As we saw in our discussion of the FAKE GUN example in chapter 19, there are natural dimensions to our categories for objects: *perceptual*, based on the conception of the object by means of our sensory apparatus; *motor activity*, based on the nature of motor interactions with objects; *functional*, based on our conception of the functions of the object; and *purposive*, based on the uses we can make of an object in a given situation. Our categories for kinds of objects are thus gestalts with at least these natural dimensions, each of which specifies interactional properties. Similarly, there are natural dimensions in terms of which we categorize events, activities, and other experiences as structured wholes. As we saw in our discussion of CONVERSATION and ARGUMENT, these natural dimensions include *participants, parts, stages, linear sequence, purpose,* and *causation*.

A categorization is a natural way of identifying a *kind* of object or experience by highlighting certain properties, downplaying others, and hiding still others. Each of the dimensions gives the properties that are highlighted. To highlight certain properties is necessarily to downplay or hide others, which is what happens whenever we categorize something. Focusing on one set of properties shifts our attention away from others. When we give everyday descriptions, for example, we are using categorizations to focus on certain properties that fit our purposes. Every description will highlight, downplay, and hide—for example:

> I've invited a sexy blonde to our dinner party.
> I've invited a renowned cellist to our dinner party.
> I've invited a Marxist to our dinner party.
> I've invited a lesbian to our dinner party.

Though the same person may fit all of these descriptions, each description highlights different aspects of the person. Describing someone who you know has all of these properties as "a sexy blonde" is to downplay the fact that she is a renowned cellist and a Marxist and to hide her lesbianism.

In general, the true statements that we make are based on the way we categorize things and, therefore, on what is highlighted by the natural dimensions of the categories. In making a statement, we make a choice of categories because we have some reason for focusing on certain properties and downplaying others. Every true statement, therefore, necessarily leaves out what is downplayed or hidden by the categories used in it.

Moreover, since the natural dimensions of categories (perceptual, functional, etc.) arise out of our interactions with the world, the properties given by those dimensions are not properties of objects *in themselves* but are, rather, interactional properties, based on the human perceptual apparatus, human conceptions of function, etc. It follows from this that true statements made in terms of human categories typically do not predicate *properties of objects in themselves* but rather *interactional properties* that make sense only relative to human functioning.

In making a true statement, we have to choose categories of description, and that choice involves our perceptions and our purposes in the given situation. Suppose you say to me, "We're having a discussion group over tonight, and I need four more chairs. Can you bring them?" I say "Sure," and show up with a hardback chair, a rocking chair, a beanbag chair, and a hassock. Leaving them in your living room, I report to you in the kitchen, "I brought the four chairs you wanted." In this situation, my statement is true, since the four objects I've brought will serve the purpose of chairs for an informal discussion group. Had you instead asked me to bring four chairs for a formal dinner and I show up with the same four objects and make the same statement, you will not be appropriately grateful and will find the statement misleading or false, since the hassock, beanbag chair, and rocker are not practical as "chairs" at a formal dinner.

This shows that our categories (e.g., CHAIR) are not rigidly fixed in terms of inherent properties of the objects themselves. What counts as an instance of a category depends on our purpose in using the category. This is the same point we made above, in our discussion of *Definition*, where we showed that categories are defined for purposes of human understanding by prototypes and family resemblances to those prototypes. Such categories are not fixed but may be narrowed, expanded, or adjusted relative to our purposes and other contextual factors. Since the truth of a statement depends on whether the categories employed in the statement fit, the truth of a statement will always be relative to the way the category is understood for our purposes in a given context.

There are many celebrated examples to show that sentences, in general, are not true or false independent of human purposes:

France is hexagonal.
Missouri is a parallelogram.
The earth is a sphere.
Italy is boot-shaped.
An atom is a tiny solar system with the nucleus at the center
and electrons whirling around it.
Light consists of particles.
Light consists of waves.

Each of these sentences is true for certain purposes, in certain respects, and in certain contexts. "France is a hexagon" and "Missouri is a parallelogram" can be true for a schoolboy who has to draw rough maps but not for professional cartographers. "The earth is a sphere" is true as far as most of us are concerned, but it won't do for precisely plotting the orbit of a satellite. No self-respecting physicist has believed since 1914 that an atom is a tiny solar system, but it is true for most of us relative to our everyday functioning and our general level of sophistication in mathematics and physics. "Light consists of particles" seems to contradict "Light consists of waves," but both are taken as true by physicists relative to which aspects of light are picked out by different experiments.

What all of this shows is that truth depends on categorization in the following four ways:

—A statement can be true only relative to some understanding of it.
—Understanding always involves human categorization, which is a function of interactional (rather than inherent) properties and of dimensions that emerge from our experience.
—The truth of a statement is always relative to the properties that are highlighted by the categories used in the statement. (For example, "Light consists of waves" highlights wavelike properties of light and hides particle-like properties.)
—Categories are neither fixed nor uniform. They are defined by prototypes and family resemblances to prototypes and are

adjustable in context, given various purposes. Whether a statement is true depends on whether the category employed in the statement fits, and this in turn varies with human purposes and other aspects of context.

What Does It Take to Understand a Simple Sentence as Being True?

To understand a sentence as being true, we must first understand it. Let us look at part of what is involved in understanding such simple sentences as "The fog is in front of the mountain" and "John fired the gun at Harry." Sentences like these are always uttered as part of discourses of some kind, and understanding them in a discourse context involves complications of a nontrivial sort that, for our purposes, we must ignore here. But, even ignoring some of the complexities of discourse context, any understanding of such sentences involves quite a bit. Consider what must be the case for us to understand "The fog is in front of the mountain" as being true. As we saw above, we have to view "the fog" and "the mountain" as entities, by means of projection, and we must project a front-back orientation on the mountain—an orientation which varies from culture to culture, is given relative to a human observer, and is not inherent in the mountain. We must then determine, relative to our purposes, whether what we view as "the fog" is pretty much between us and what we pick out as "the mountain," closer to the mountain, and not to the side of the mountain, or above it, etc. There are three projections onto the world plus some pragmatic determinations, relative to our perceptions and purposes, as to whether the relation *in front of* is more appropriate than other possible relations. Thus, understanding whether "The fog is in front of the mountain" is true is not merely a matter of (*a*) picking out preexisting and well-defined entities in the world (the fog and the mountain) and (*b*) seeing whether some inherent relation (independent of any human observer) holds between these well-defined entities. Instead, it is a matter of human projection and human judgment, relative to certain purposes.

"John fired the gun at Harry" raises other issues. There are the obvious matters of picking out people named *John* and *Harry,* picking out the object that fits the category GUN, understanding what it means to fire a gun and to fire it at someone. But we don't understand sentences like this *in vacuo.* We understand them relative to certain larger categories of experience, for example, shooting someone, scaring someone, performing a circus act, or pretending to do any of these in a play or film or joke. Firing a gun can be an instance of any of these, and which is applicable will depend on the context. But there is only a small range of categories of experience that firing a gun fits into, the most typical of which is SHOOTING SOMEONE, since there are many typical ways to scare someone or perform a circus act but only one normal way to shoot someone.

We can thus view SHOOTING SOMEONE as an experiential gestalt with roughly the following dimensions, in this instance:

Participants:	John (*shooter*), Harry (*target*), the gun (*instrument*), the bullet (*instrument, missile*)
Parts:	Aiming the gun at the target Firing the gun Bullet hits target Target is wounded
Stages:	*Precondition:* Shooter has loaded gun *Beginning:* Shooter aims gun at target *Middle:* Shooter fires gun *End:* Bullet hits target *Final state:* Target is wounded
Causation:	Beginning and middle enable end Middle and end cause final state
Purpose:	*Goal:* Final state *Plan:* Meet precondition, perform beginning and middle

The sentence "John fired the gun at Harry" typically

evokes a SHOOTING SOMEONE gestalt of this form. Or it could, in other contexts, evoke other equally complex experiential gestalts (e.g., PERFORMING A CIRCUS ACT). But the sentence is virtually never understood on its own terms without the evocation of some larger gestalt that specifies the normal range of natural dimensions (e.g., purpose, stages, etc.). Whichever gestalt is evoked, we understand much more than is given directly in the sentence. Each such gestalt provides a background for understanding the sentence in terms that make sense to us, that is, in terms of an experiential category of our culture.

In addition to the larger category of experience evoked by the sentence, we also categorize FIRING and GUN in terms of information-rich prototypes. Unless the context forces us to do otherwise, we understand the gun to be a prototypical gun, with the usual prototypical *perceptual, motor, functional,* and *purpose* properties. Unless the context specifies otherwise, the image evoked is not that of an umbrella gun or a suitcase gun, and the motor program used in firing is holding the gun horizontal and squeezing the trigger, which is the normal motor program that fits both FIRING and GUN. Unless the context is rigged, we do not imagine a Rube Goldberg device in which the trigger is tied by a string to, say, a door handle.

We understand the sentence in terms of the way these gestalts fit together, both the "smaller" gestalts (GUN, FIRING, AIMING, etc.) and the "larger" gestalts (SHOOTING SOMEONE or PERFORMING A CIRCUS ACT). Only relative to such understandings do issues of truth arise. The issue of truth is straightforward when our understanding of the sentence in these terms fits closely enough our understanding of the events that have occurred. But what happens when there is a discrepancy between our *normal* understanding of the sentence and our understanding of the events? Say, for example, that John, in an ingenious Rube Goldberg fashion, set up the gun so that it would be aimed at a point where Harry would be at some time and then tied a string to the trigger. Let's take two cases:

A. John's scratching his ear causes the gun to fire at Harry.
B. Harry's opening the door causes the gun to fire at Harry.

In case A, John's action is responsible for the firing, while, in B, Harry's action is. This makes A closer than B to our normal understanding of the sentence. Thus, we might, if pressed, be willing to say that A is a case where it would be true to say "John fired the gun at Harry." Case B, however, is so far from our prototypical understanding of firing that we would probably not want to say that it was true that "John fired the gun at Harry." But we would not want to say that it was unqualifiedly false either, since John was primarily responsible for the shooting. Instead, we'd want to explain, not just answer "True" or "False." This is what typically happens when our understanding of the events does not fit our normal understanding of the sentence because of some deviation from a prototype.

We can summarize the results of this section as follows:

1. Understanding a sentence as being true in a given situation requires having an understanding of the sentence and having an understanding of the situation.
2. We understand a sentence as being true when our understanding of the sentence fits our understanding of the situation closely enough.
3. Getting an understanding of a situation of the sort that could fit our understanding of a sentence may require:
   a. Projecting an orientation onto something that has no inherent orientation (e.g., viewing the mountain as having a front)
   b. Projecting an entity structure onto something that is not bounded in any clear sense (e.g., the fog, the mountain)
   c. Providing a background in terms of which the sentence makes sense, that is, calling up an experiential gestalt (e.g., SHOOTING SOMEONE, PERFORMING A CIRCUS ACT) and understanding the situation in terms of the gestalt
   d. Getting a "normal" understanding of the sentence in terms of its categories (e.g., GUN, FIRING), as defined by prototype, and trying to get an understanding of the situation in terms of the same categories

What Does It Take to Understand a
Conventional Metaphor as Being True?

We have seen what is involved in understanding a simple sentence (without metaphor) as being true. We now want to suggest that adding conventional metaphors changes nothing. We understand them as being true in basically the same way. Take a sentence like "Inflation has gone up." Understanding a situation as one in which this sentence could be true involves two projections. We have to pick out instances of inflation and view them as constituting a substance, which we can then quantify and thereby view as increasing. In addition we have to project an UP orientation on the increase. These two projections constitute two conventional metaphors: INFLATION IS A SUBSTANCE (an ontological metaphor) and MORE IS UP (an orientational metaphor). There is one principal difference between the projections onto the situation in this case and in the case given above, namely, "The fog is in front of the mountain." In the case of *fog*, we are understanding something physical (fog) on the model of something else physical but more clearly delineated—a bounded physical object. In the case of *front*, we are understanding the physical orientation of the mountain in terms of another physical orientation—that of our bodies. In both cases, we are understanding something that is physical in terms of something *else* that is *also* physical. In other words, we are understanding one thing in terms of something else *of the same kind*. But in conventional metaphor, we are understanding one thing in terms of something else *of a different kind*. In "Inflation has gone up," for example, we understand *inflation* (which is abstract) in terms of a physical substance, and we understand an increase of inflation (which is also abstract) in terms of a physical orientation (up). The only difference is whether our projection involves the *same* kinds of things or *different* kinds of things.

When we understand a sentence like "Inflation has gone up" as being true, we do the following:

1. We understand the *situation* by metaphorical projection in two ways:
   a. We view inflation as a SUBSTANCE (via an ontological metaphor).
   b. We view MORE as being oriented UP (via an orientational metaphor).
2. We understand the *sentence* in terms of the same two metaphors.
3. This allows us to fit our understanding of the sentence to our understanding of the situation.

Thus an understanding of truth in terms of metaphorical projection is not essentially different from an understanding of truth in terms of nonmetaphorical projection. The only difference is that metaphorical projection involves understanding one kind of thing in terms of another kind of thing. That is, metaphorical projection involves two different kinds of things, while nonmetaphorical projection involves only one kind.

The same holds for structural metaphors, also. Take a sentence like "John defended his position in the argument." As we saw above, the experience of arguing is structured partially in terms of the WAR gestalt, by virtue of the ARGUMENT IS WAR metaphor. Since argument is a metaphorical kind of experience, structured by the conventional metaphor ARGUMENT IS WAR, it follows that a situation in which there is an argument may be understood in these metaphorical terms. Our understanding of an argument situation will involve viewing it simultaneously in terms of both the CONVERSATION gestalt and the WAR gestalt. If our understanding of the situation is such that a portion of the conversation fits a successful defense in the WAR gestalt, then our understanding of the sentence will fit our understanding of the situation and we will take the sentence to be true.

In both the metaphorical and nonmetaphorical cases, our account of how we understand truth depends on our account of how we understand situations. Given that metaphor is conceptual in nature rather than a matter of "mere language," it is natural for us to conceptualize situations in metaphorical terms. Because we can conceptualize *situations* in metaphorical terms, it is possible for *sentences* containing metaphors to be taken as fitting the situations as we conceptualize them.

How Do We Understand New Metaphors as Being True?

We have just seen that conventional metaphors fit our account of truth in the same way as nonmetaphorical sentences do. In both cases, understanding a sentence as true in a given situation involves fitting our understanding of the sentence to our understanding of the situation. Because our understanding of situations may involve conventional metaphor, sentences with conventional metaphors raise no special problems for our account of truth. This suggests that the same account of truth should work for new, or nonconventional, metaphors.

To see this, let us consider two related metaphors, one conventional and one nonconventional:

Tell me the *story of your life*. (conventional)
Life's . . . a tale told by an idiot, full of sound and fury, signifying nothing. (nonconventional)

Let us start with "Tell me the story of your life," which contains the conventional metaphor LIFE IS A STORY. This is a metaphor rooted deep in our culture. It is assumed that everyone's life is structured like a story, and the entire biographical and autobiographical tradition is based on this assumption. Suppose someone asks you to tell your life story. What do you do? You construct a coherent narrative that starts early in your life and continues up to the present. Typically the narrative will have the following features:

Participants:	You and other people who have "played a role" in your life
Parts:	Settings, significant facts, episodes, and

	significant states (including the present state and some original state)
Stages:	*Preconditions:* Setting for the *beginning* *Beginning:* The original state followed by episodes in the same temporal setting *Middle:* Various episodes and significant states, in succeeding temporal order. *End:* Present state
Linear sequence:	Various temporal and/or causal connections among the succeeding episodes and states
Causation:	Various causal relations between episodes and states
Purpose:	*Goal:* A desired state (which may be in the future) *Plan:* A sequence of episodes which you initiate and which have a causal connection to the goal or: An event or set of events that puts you in a significant state, so that you will reach the goal through a series of natural stages

This is a much oversimplified version of a typical experiential gestalt for giving coherence to one's life by viewing it as a STORY. We have omitted various complexities, such as the fact that each episode may in itself be a coherent subnarrative with a similar structure. Not all life stories will contain all of these dimensions of structure.

Notice that understanding your life in terms of a coherent life story involves highlighting certain *participants* and *parts* (episodes and states) and ignoring or hiding others. It involves seeing your life in terms of *stages, causal connections* among the parts, and *plans* meant to achieve a *goal* or a set of goals. In general, a life story imposes a coherent structure on elements of your life that are highlighted.

If you tell such a story and then say, "That is the story of my life," you will legitimately see yourself as telling the truth if you do, in fact, view the highlighted participants and events as the significant ones and do, in fact, perceive them as fitting together coherently in the way specified by the structure of the narrative. The issue of truth in this case is whether the coherence provided by the narrative matches the coherence you see in your life. And it is the coherence that you see in your life that gives it meaning and significance.

Let us now ask what is involved in understanding as true the nonconventional metaphor "Life's . . . a tale told by an idiot, full of sound and fury, signifying nothing." This non-conventional metaphor evokes the conventional metaphor LIFE IS A STORY. The most salient fact about stories told by idiots is that they are not coherent. They start off as if they were coherent stories with stages, causal connections, and overall purposes, but they suddenly shift over and over again, making it impossible to find coherence as you go along or any coherence overall. A life story of this sort would have no coherent structure for us and therefore no way of providing meaning or significance to our lives. There would be no way of highlighting events in your life as being significant, that is, as serving a purpose, having a causal connection to other significant events, fitting stages, etc. In a life viewed as a tale, episodes "full of sound and fury" would represent periods of frenzy, agonized struggle, and perhaps violence. In a typical life story, such events would be viewed as momentous—either traumatic, cathartic, dis-astrous, or climactic. But the modifier "signifying nothing" negates all these possibilities for significance, suggesting instead that the episodes cannot be viewed in terms of causal connections, purposes, or identifiable stages in some coherent whole.

If we in fact view our lives and the lives of others in this way, then we would take the metaphor as being true. What makes it possible for many of us to see this metaphor as true is that we usually comprehend our life experiences in terms of the LIFE IS A STORY metaphor. We are constantly looking for meaning in our lives by seeking out coherences that will fit some sort of coherent life story. And we constantly tell such stories and live in terms of them. As the circumstances of our lives change, we constantly revise our life stories, seeking new coherence.

The metaphor LIFE'S . . . A TALE TOLD BY AN IDIOT may well fit the lives of people whose life circumstances change so radically, rapidly, and unexpectedly that no coherent life story ever seems possible for them.

Although we have seen that such new, nonconventional metaphors will fit our general account of truth, we should stress again that issues of truth are among the least relevant and interesting issues that arise in the study of metaphor. The real significance of the metaphor LIFE'S . . . A TALE TOLD BY AN IDIOT is that, in getting us to try to understand how it could be true, it makes possible a new understanding of our lives. It highlights the fact that we are constantly functioning under the expectation of being able to fit our lives into some coherent life story but that this expectation may be constantly frustrated when the most salient experiences in our lives, those full of sound and fury, do not fit any coherent whole and, therefore, signify nothing. Normally, when we construct life stories, we leave out many extremely important experiences for the sake of finding coherence. What the LIFE'S . . . A TALE TOLD BY AN IDIOT metaphor does is to evoke the LIFE IS A STORY metaphor, which involves living with the constant expectation of fitting important episodes into a coherent whole—a sane life story. The effect of the metaphor is to evoke this expectation and to point out that, in reality, it may be constantly frustrated.

Understanding a Situation: A Summary

In this chapter we have been developing the elements of an experiential account of truth. Our account of truth is based on understanding. What is central to this theory is our analysis of what it means to understand a situation. Here is a summary of what we have said on the matter so far:

Direct Immediate Understanding

There are many things that we understand directly from our direct physical involvement as an inseparable part of our immediate environment.

Entity structure: We understand ourselves as bounded entities, and we directly experience certain objects that we come into direct contact with as bounded entities, too.

Orientational structure: We understand ourselves and other objects as having certain orientations relative to the environments we function in (up-down, in-out, front-back, on-off, etc.).

Dimensions of experience: There are dimensions of experience in terms of which we function most of the time in our direct interactions with others and with our immediate physical and cultural environment. We categorize the entities we directly encounter and the direct experiences we have in terms of these categories.

Experiential gestalts: Our object and substance categories are gestalts that have at least the following dimensions: *perceptual, motor activity, part/whole, functional, purposive.* Our categories of direct actions, activities, events, and experiences are gestalts that have at least the following dimensions: *participants, parts, motor activities, perceptions, stages, linear sequences (of parts), causal relations, purpose (goals/plans* for actions and *end states* for events). These constitute the natural dimensions of our direct experience. Not all of them will play a role in every kind of direct experience, but, in general, most of them will play some role or other.

Background: An experiential gestalt will typically serve as a background for understanding something we experience as an aspect of that gestalt. Thus a person or object may be understood as a *participant* in a gestalt, and an action may be understood as a *part* of a gestalt. One gestalt may presuppose the presence of another, which may, in turn, presuppose the presence of others, and so on. The result will typically be an incredibly rich background structure necessary for a full understanding of any given situation. Most of this background structure will never be noticed, since it is presupposed in so many of our daily activities and experiences.

Highlighting: Understanding a situation as being an instance of an experiential gestalt involves picking out elements of the situation as fitting the dimensions of the gestalt—for example, picking out aspects of the experience as being *participants, parts, stages,* etc. This highlights those aspects of the situation and downplays or hides aspects of the situation that do not fit the gestalt.

Interactional properties: The properties we directly experience an object or event as having are products of our interactions with them in our environment. That is, they may not be *inherent* properties of the object or experience but, instead, *interactional* properties.

Prototypes: Each category is structured in terms of a prototype, and something counts as a member of the category by virtue of the family resemblances it bears to the prototype.

Indirect Understanding

We have just described how we understand aspects of a situation that are fairly clearly delineated in our direct experience. But we have seen throughout this work that many aspects of our experience cannot be clearly delineated in terms of the naturally emergent dimensions of our experience. This is typically the case for human emotions, abstract concepts, mental activity, time, work, human institutions, social practices, etc., and even for physical objects that have no inherent boundaries or orientations. Though most of these can be *experienced* directly, none of them can be fully comprehended on their own terms. Instead, we must understand them in terms of other entities and experiences, typically other *kinds* of entities and experiences.

As we saw, understanding a situation where we see the fog as being in front of the mountain requires us to view the fog and the mountain as entities. It also requires us to project a front-back orientation upon the mountain. These projections are built into our very perception. We perceive the fog and the mountain as entities and we perceive the mountain as having a front, with the fog in front of it. The front-back orientation that we perceive for the mountain is obviously an interactional property, as is the status of the fog and the mountain as entities. Here we have a case of indirect understanding, where we are understanding physical phenomena in terms of other more clearly delineated physical phenomena.

What we do in indirect understanding is to use the resources of direct understanding. In the case of the fog and the mountain, we are using entity structure and orientational structure. In this case we stayed within a single domain, that of physical objects. But most of our indirect understanding involves understanding *one kind* of entity or experience in terms of *another kind*—that is, understanding via metaphor. As we have seen, all of the resources that are used in direct, immediate understanding are pressed into service in indirect understanding via metaphor.

Entity structure: Entity and substance structure is imposed via ontological metaphor.

Orientational structure: Orientational structure is imposed via orientational metaphor.

Dimensions of experience: Structural metaphor involves structuring one kind of thing or experience in terms of another kind, but the same natural dimensions of experience are used in both (e.g., *parts, stages, purposes,* etc.).

Experiential gestalts: Structural metaphor involves imposing part of one gestalt structure upon another.

Background: Experiential gestalts play the role of a background in metaphorical understanding, just as they do in nonmetaphorical understanding.

Highlighting: Metaphorical highlighting works by the same mechanism as that for nonmetaphorical gestalts. That is, the experiential gestalt that is superimposed in the situation via the metaphor picks out elements of the situation as fitting its dimensions—it picks out its own participants, parts, stages, etc. These are what the metaphor highlights, and what is not highlighted is downplayed or hidden.

Since new metaphors highlight things not usually highlighted by our normal conceptual structure, they have become the most celebrated examples of highlighting.

Interactional properties: All of the dimensions of our experience are interactional in nature, and all experiential gestalts involve interactional properties. This holds for both metaphorical and nonmetaphorical concepts.

Prototypes: Both metaphorical and nonmetaphorical categories are structured in terms of prototypes.

Truth Is Based on Understanding

We have seen that the same eight aspects of our conceptual system that go into direct immediate understanding of situations play parallel roles in indirect understanding. These aspects of our normal conceptual system are used whether we are understanding a situation in metaphorical or nonmetaphorical terms. It is because we understand *situations* in terms of our conceptual system that we can understand *statements* using that system of concepts as being *true,* that is, as fitting or not fitting the situation as we understand it. Truth is therefore a function of our conceptual system. It is because many of our concepts are metaphorical in nature, and because we understand situations in terms of those concepts, that metaphors can be true or false.

The Nature of the Experientialist Account of Truth

We understand a statement as being true in a given situation when our understanding of the statement fits our understanding of the situation closely enough for our purposes.

This is the foundation of our experientialist theory of truth, which has the following characteristics.

First, our theory has some elements in common with a *correspondence* theory. According to the most rudimentary correspondence view, a statement has an objective meaning, which specifies the conditions under which it is true. Truth consists of a direct fit (or correspondence) between a statement and some state of affairs in the world.

We reject such a simplistic picture, primarily because it ignores the way in which truth is based on understanding. The experientialist view we are proposing is a correspondence theory in the following sense:

A theory of truth is a theory of what it means to understand a statement as true or false in a certain situation.

Any correspondence between what we say and some state of affairs in the world is always mediated by our understanding of the statement and of the state of affairs. Of course, our understanding of the situation results from our interaction with the situation itself. But we are able to make true (or false) statements about the world because it is possible for *our understanding of a statement* to fit (or not fit) *our understanding of the situation* in which the statement is made.

Since we understand situations and statements in terms of our conceptual system, truth for us is always relative to that conceptual system. Likewise, since an understanding is always partial, we have no access to "the whole truth" or to any definitive account of reality.

Second, understanding something requires fitting it into a coherent scheme, relative to a conceptual system. Thus, truth will always depend partly on coherence. This gives us elements of a *coherence theory*.

Third, understanding also requires a grounding in experience. On the experientialist view, our conceptual system emerges from our constant successful functioning in our physical and cultural environment. Our categories of experience and the dimensions out of which they are constructed not only have emerged from our experience but are constantly being tested through ongoing successful functioning by all the members of our culture. This gives us elements of a *pragmatic theory*.

Fourth, the experientialist theory of truth has some elements in common with classical *realism*, but these do not include its insistence on absolute truth. Instead, it takes as given that:

> The physical world is what it is. Cultures are what they are. People are what they are.
>
> People successfully interact in their physical and cultural environments. They are constantly interacting with the real world.
>
> Human categorization is constrained by reality, since it is characterized in terms of natural dimensions of experience that are constantly tested through physical and cultural interaction.
>
> Classical realism focuses on physical reality rather than cultural and personal reality. But social, political, economic, and religious institutions and the human beings who function within them are no less real than trees, tables, or rocks. Since our account of truth deals with social and personal reality as well as physical reality, it can be considered an attempt to extend the realist tradition.
>
> The experientialist theory varies from classical objective realism in the following basic way: Human concepts do not correspond to inherent properties of things but only to interactional properties. This is natural, since concepts can be metaphorical in nature and can vary from culture to culture.

Fifth, people with very different conceptual systems than our own may understand the world in a very different way than we do. Thus, they may have a very different body of truths than we have and even different criteria for truth and reality.

It should be obvious from this description that there is nothing radically new in our account of truth. It includes some of the central insights of the phenomenological tradition, such as the rejection of epistemological foundationalism, the stress on the centrality of the body in the structuring of our experience, and the importance of that structure in understanding. Our view also accords with some of the key elements of Wittgenstein's later philosophy: the family-resemblance account of categorization, the rejection of the picture theory of meaning, the rejection of a building-block theory of meaning, and the emphasis on meaning as relative to context and to one's own conceptual system.

Elements of Human Understanding in Theories of "Objective Truth"

A theory of truth based on understanding is obviously not a theory of "purely objective truth." We do not believe that there is such a thing as absolute truth, and we think that it is pointless to try to give a theory of it. However, it is traditional in Western philosophy to assume that absolute truth is possible and to undertake to give an account of it. We would like to point out how the most prominent contemporary approaches to the problem build in aspects of human understanding, which they claim to exclude.

The most obvious case is the account of truth given within model-theoretic approaches, say, for example, those done within the Kripke and Montague traditions. The models are constructed out of a universe of discourse that is taken to be a set of *entities*. Relative to this set of entities, we can define world states, in which all the properties that the entities have and all the relations among them are specified. It is assumed that this concept of a world state is sufficiently general to apply to any conceivable situation, including the real world. In such a system, sentences like "The fog is in front of the mountain" would present no problem, since there would be an entity corresponding to *the fog*, an entity corresponding to *the mountain*, and a relation *in front of*, relating the two entities. But such models do not correspond to the world in itself, free of human understanding, since there are in the world no well-defined entities corresponding to *the mountain* and *the fog* and there is no inherent *front* to the mountain. The entity structure and the front-back orientation are imposed by virtue of human understanding. Any attempt to give an account of the truth of "The fog is in front of the mountain" in such model-theoretic terms will not be an account of *objective, absolute* truth, since it involves building elements of human understanding into the models.

The same can be said of attempts to provide a theory of truth meeting the constraints of the classic Tarski truth definition:

> "*S*" is true if and only if *S* . . .

or more up-to-date versions like:

> "*S*" is true if and only if *p* (where *p* is a statement in some universally applicable logical language)

The prototype for such theories, the well-worn

> "Snow is white" is true if and only if snow is white.

seems reasonable enough, since there could reasonably be thought to be a sense in which snow is objectively identifiable and in which it is inherently white. But what about

> "The fog is in front of the mountain" is true if and only if the fog is in front of the mountain.

Since the world does not contain clearly identifiable entities *the fog* and *the mountain,* and since mountains don't have inherent fronts, the theory can work only relative to some human understanding of what a front is for a mountain and to some delineation of *fog* and *mountain.* The problem is even trickier, since not all human beings have the same way of projecting fronts onto mountains. Here some elements of human understanding must be brought in to make the truth definition work.

There is another important difference between our account of truth in terms of understanding and the standard attempts to give an account of truth free of human understanding. The different accounts of truth give rise to different accounts of meaning. For us, meaning depends on understanding. A sentence can't mean anything to you unless you understand it. Moreover, meaning is always meaning *to* someone. There is no such thing as a meaning of a sentence in itself, independent of any people. When we speak of the meaning of a sentence, it is always the meaning of the sentence to someone, a real person or a hypothetical typical member of a speech community.

Here our theory differs radically from standard theories of meaning. The standard theories assume that it is possible to give an account of truth in itself, free of human understanding, and that the theory of meaning will be based on such a theory of truth. We see no possibility for any such program to work and think that the only answer is to base both the theory of meaning and the theory of truth on a theory of understanding. Metaphor, both conventional and nonconventional, plays a central role in such a program. Metaphors are basically devices for understanding and have little to do with objective reality, if there is such a thing. The fact that our conceptual system is inherently metaphorical, the fact that we understand the world, think, and function in metaphorical terms, and the fact that metaphors can not merely be understood but can be meaningful and true as well—these facts all suggest that an adequate account of meaning and truth can only be based on understanding.

25

The Myths of Objectivism and Subjectivism

The Choices Our Culture Offers

We have given an account of the way in which truth is based on understanding. We have argued that truth is always relative to a conceptual system, that any human conceptual system is mostly metaphorical in nature, and that, therefore, there is no fully objective, unconditional, or absolute truth. To many people raised in the culture of sci ence or in other subcultures where absolute truth is taken for granted, this will be seen as a surrender to subjectivity and arbitrariness—to the Humpty-Dumpty notion that something means "just what I choose it to mean—neither more nor less." For the same reason, those who identify with the Romantic tradition may see any victory over objectivism as a triumph of imagination over science—a triumph of the view that each individual makes his own reality, free of any constraints.

Either of these views would be a misunderstanding based on the mistaken cultural assumption that the only alternative to objectivism is radical subjectivity—that is, either you believe in absolute truth *or* you can make the world in your own image. If you're not being *objective,* you're being *subjective,* and there is no third choice. We see ourselves as offering a third choice to the myths of objectivism and subjectivism.

Incidentally, we are not using the term "myth" in any derogatory way. Myths provide ways of comprehending experience; they give order to our lives. Like metaphors, myths are necessary for making sense of what goes on around us. All cultures have myths, and people cannot function without myth any more than they can function without metaphor. And just as we often take the *metaphors* of our own culture as truths, so we often take the *myths* of our own culture as truths. The myth of objectivism is particularly insidious in this way. Not only does it purport not to be a myth, but it makes both myths and metaphors objects of belittlement and scorn: according to the objectivist myth, myths and metaphors cannot be taken seriously because they are not objectively true. As we will see, the myth of objectivism is itself not objectively true. But this does not make it something to be scorned or ridiculed. The myth of objectivism is part of the everyday functioning of every member of this culture. It needs to be examined and understood. We also think it needs to be supplemented—not by its opposite, the myth of subjectivism, but by a new experientialist myth, which we think better fits the realities of our experience. In order to get clear about what an experientialist alternative would be like, we first need to examine the myths of objectivism and subjectivism in detail.

The Myth of Objectivism

The myth of objectivism says that:

1. The world is made up of objects. They have properties independent of any people or other beings who experience them. For example, take a rock. It's a separate object and it's hard. Even if no people or other beings existed in the universe, it would still be a separate object and it would still be hard.

2. We get our knowledge of the world by experiencing the objects in it and getting to know what properties the objects have and how these objects are related to one another. For example, we find out that a rock is a separate object by looking at it, feeling it, moving it around, etc. We find out that it's hard by touching it, trying to squeeze it, kicking it, banging it against something softer, etc.

3. We understand the objects in our world in terms of categories and concepts. These categories and concepts correspond to properties the objects have in themselves (inherently) and to the relationships among the objects. Thus, we have a word "rock," which corresponds to a concept ROCK. Given a rock, we can tell that it is in the category ROCK and that a piano, a tree, or a tiger would not be. Rocks have inherent properties independent of any beings: they're solid, hard, dense, occur in nature, etc. We understand what a "rock" is in terms of these properties.

4. There is an objective reality, and we can say things that are objectively, absolutely, and unconditionally true and false about it. But, as human beings, we are subject to human error, that is, illusions, errors of perception, errors of judgment, emotions, and personal and cultural biases. We cannot rely upon the subjective judgments of individual people. Science provides us with a methodology that allows us to rise above our subjective limitations and to achieve understanding from a universally valid and unbiased point of view. Science can ultimately give a correct, definitive, and general account of reality, and, through its methodology, it is constantly progressing toward that goal.

5. Words have fixed meanings. That is, our language expresses the concepts and categories that we think in terms of. To describe reality correctly, we need words whose meanings are clear and precise, words that fit reality. These may be words that arise naturally, or they may be technical terms in a scientific theory.

6. People can be objective and can speak objectively, but they can do so only if they use language that is clearly and precisely defined, that is straightforward and direct, and that can fit reality. Only by speaking in this way can people communicate precisely about the external world and make statements that can be judged objectively to be true or false.

7. Metaphor and other kinds of poetic, fanciful, rhetorical, or figurative language can always be avoided in speaking objectively, and they should be avoided, since their

meanings are not clear and precise and do not fit reality in any obvious way.

8. Being objective is generally a good thing. Only objective knowledge is really knowledge. Only from an objective, unconditional point of view can we really understand ourselves, others, and the external world. Objectivity allows us to rise above personal prejudice and bias, to be fair, and to take an unbiased view of the world.

9. To be objective is to be rational; to be subjective is to be irrational and to give in to the emotions.

10. Subjectivity can be dangerous, since it can lead to losing touch with reality. Subjectivity can be unfair, since it takes a personal point of view and can, therefore, be biased. Subjectivity is self-indulgent, since it exaggerates the importance of the individual.

The Myth of Subjectivism

The myth of subjectivism says that:

1. In most of our everyday practical activities we rely on our senses and develop intuitions we can trust. When important issues arise, regardless of what others may say, our own senses and intuitions are our best guides for action.

2. The most important things in our lives are our feelings, aesthetic sensibilities, moral practices, and spiritual awareness. These are purely subjective. None of these is purely rational or objective.

3. Art and poetry transcend rationality and objectivity and put us in touch with the more important reality of our feelings and intuitions. We gain this awareness through imagination rather than reason.

4. The language of the imagination, especially metaphor, is necessary for expressing the unique and most personally significant aspects of our experience. In matters of personal understanding the ordinary agreed-upon meanings that words have will not do.

5. Objectivity can be dangerous, because it misses what

is most important and meaningful to individual people. Objectivity can be unfair, since it must ignore the most relevant realms of our experience in favor of the abstract, universal, and impersonal. For the same reason, objectivity can be inhuman. There are no objective and rational means for getting at our feelings, our aesthetic sensibilities, etc. Science is of no use when it comes to the most important things in our lives.

Fear of Metaphor

Objectivism and subjectivism need each other in order to exist. Each defines itself in opposition to the other and sees the other as the enemy. Objectivism takes as its allies scientific truth, rationality, precision, fairness, and impartiality. Subjectivism takes as its allies the emotions, intuitive insight, imagination, humaneness, art, and a "higher" truth. Each is master in its own realm and views its realm as the better of the two. They coexist, but in separate domains. Each of us has a realm in his life where it is appropriate to be objective and a realm where it is appropriate to be subjective. The portions of our lives governed by objectivism and subjectivism vary greatly from person to person and from culture to culture. Some of us even attempt to live our entire lives totally by one myth or the other.

In Western culture as a whole, objectivism is by far the greater potentate, claiming to rule, at least nominally, the realms of science, law, government, journalism, morality, business, economics, and scholarship. But, as we have argued, objectivism is a myth.

Since the time of the Greeks, there has been in Western culture a tension between truth, on the one hand, and art, on the other, with art viewed as illusion and allied, via its link with poetry and theater, to the tradition of persuasive public oratory. Plato viewed poetry and rhetoric with suspicion and banned poetry from his utopian Republic because it gives no truth of its own, stirs up the emotions, and thereby blinds mankind to the real truth. Plato, typical of persuasive writers, stated his view that truth is absolute and art mere illusion by the use of a powerful rhetorical device, his Allegory of the Cave. To this day, his metaphors dominate Western philosophy, providing subtle and elegant expression for his view that truth is absolute. Aristotle, on the other hand, saw poetry as having a positive value: "It is a great thing, indeed, to make proper use of the poetic forms, . . . But the greatest thing by far is to be a master of metaphor" (*Poetics* 1459a); "ordinary words convey only what we know already; it is from metaphor that we can best get hold of something fresh" (*Rhetoric* 1410b).

But although Aristotle's theory of how metaphors work is *the* classic view, his praise of metaphor's ability to induce insight was never carried over into modern philosophical thought. With the rise of empirical science as a model for truth, the suspicion of poetry and rhetoric became dominant in Western thought, with metaphor and other figurative devices becoming objects of scorn once again. Hobbes, for example, finds metaphors absurd and misleadingly emotional; they are *"ignes fatui;* and reasoning upon them is wandering amongst innumerable absurdities; and their end, contention and sedition, or contempt" (*Leviathan,* pt. 1, chap. 5). Hobbes finds absurdity in "the use of metaphors, tropes, and other rhetorical figures, instead of words proper. For though it be lawful to say, for example in common speech, *the way goeth, or leadeth hither, or thither; the proverb says this or that,* whereas ways cannot go, nor proverbs speak; yet in reckoning, and seeking of truth, such speeches are not to be admitted" (ibid.).

Locke, continuing the empiricist tradition, shows the same contempt for figurative speech, which he views as a tool of rhetoric and an enemy of truth:

> . . . if we would speak of things as they are, we must allow that all the art of rhetoric, besides order and clearness; all the artifi-

cial and figurative application of words eloquence hath invented, are for nothing else but to insinuate wrong ideas, move the passions, and thereby mislead the judgment; and so indeed are perfect cheats: and therefore, however laudable or allowable oratory may render them in harangues and popular addresses, they are certainly, in all discourses that pretend to inform or instruct, wholly to be avoided; and where truth and knowledge are concerned, cannot but be thought a great fault, either of the language or person that makes use of them. . . . It is evident how much men love to deceive and be deceived, since rhetoric, that powerful instrument of error and deceit, has its established professors, is publicly taught, and has always been had in great reputation. [*Essay Concerning Human Understanding*, bk. 3, chap. 10]

The fear of metaphor and rhetoric in the empiricist tradition is a fear of subjectivism—a fear of emotion and the imagination. Words are viewed as having "proper senses" in terms of which truths can be expressed. To use words metaphorically is to use them in an improper sense, to stir the imagination and thereby the emotions and thus to lead us away from the truth and toward illusion. The empiricist distrust and fear of metaphor is wonderfully summed up by Samuel Parker:

All those Theories in Philosophy which are expressed only in metaphorical Termes, are not real Truths, but the meer products of Imagination, dress'd up (like Childrens babies) in a few spangled empty words. . . . Thus their wanton and luxuriant fancies climbing up into the Bed of Reason, do not only defile it by unchaste and illegitimate Embraces, but instead of real conceptions and notices of Things, impregnate the mind with nothing but Ayerie and Subventaneous Phantasmes. [*Free and Impartial Censure of the Platonick Philosophy* (1666)]

As science became more powerful via technology and the Industrial Revolution became a dehumanizing reality, there occurred a reaction among poets, artists, and occasional philosophers: the development of the Romantic tradition. Wordsworth and Coleridge gladly left reason, science, and objectivity to the dehumanized empiricists and exalted imagination as a more humane means of achieving a higher truth, with emotion as a natural guide to self-understanding. Science, reason, and technology had alienated man from himself and his natural environment, or so the Romantics alleged; they saw poetry, art, and a return to nature as a way for man to recover his lost humanity. Art and poetry were seen, not as products of reason, but as "the spontaneous overflow of powerful feelings." The result of this Romantic view was the alienation of the artist and poet from mainstream society.

The Romantic tradition, by embracing subjectivism, reinforced the dichotomy between truth and reason, on the one hand, and art and imagination, on the other. By giving up on rationality, the Romantics played into the hands of the myth of objectivism, whose power has continued to increase ever since. The Romantics did, however, create a domain for themselves, where subjectivism continues to hold sway. It is an impoverished domain compared to that of objectivism. In terms of real power in our society—in science, law, government, business, and the media—the myth of objectivism reigns supreme. Subjectivism has carved out a domain for itself in art and perhaps in religion. Most people in this culture see it as an appendage to the realm of objectivism and a retreat for the emotions and the imagination.

The Third Choice: An Experientialist Synthesis

What we are offering in the experientialist account of understanding and truth is an alternative which denies that subjectivity and objectivity are our only choices. We reject the objectivist view that there is absolute and unconditional truth without adopting the subjectivist alternative of truth as obtainable only through the imagination, unconstrained by external circumstances. The reason we have focused so much on metaphor is that it unites reason and imagination. Reason, at the very least, involves categorization, entailment, and inference. Imagination, in one of its many aspects, involves seeing one kind of thing in terms of another kind of thing—what we have called metaphorical thought. Metaphor is thus *imaginative rationality*. Since the categories of our everyday thought are largely metaphorical and our everyday reasoning involves metaphorical entailments and inferences, ordinary rationality is therefore imaginative by its very nature. Given our understanding of poetic metaphor in terms of metaphorical entailments and inferences, we can see that the products of the poetic imagination are, for the same reason, partially rational in nature.

Metaphor is one of our most important tools for trying to comprehend partially what cannot be comprehended totally: our feelings, aesthetic experiences, moral practices, and spiritual awareness. These endeavors of the imagination are not devoid of rationality; since they use metaphor, they employ an imaginative rationality.

An experientialist approach also allows us to bridge the gap between the objectivist and subjectivist myths about impartiality and the possibility of being fair and objective. The two choices offered by the myths are absolute objectivity, on the one hand, and purely subjective intuition, on the other. We have seen that truth is relative to understanding, which means that there is no absolute standpoint from which to obtain absolute objective truths about the world. This does not mean that there are no truths; it means only that truth is relative to our conceptual system, which is grounded in, and constantly tested by, our experiences and those of other members of our culture in our daily interactions with other people and with our physical and cultural environments.

Though there is no absolute objectivity, there can be a kind of objectivity relative to the conceptual system of a culture. The point of impartiality and fairness in social matters is to rise above relevant *individual* biases. The point of objectivity in scientific experimentation is to factor out the effects of *individual* illusion and error. This is not to say that we can always, or even ever, be completely successful in factoring out individual biases to achieve complete objectivity relative to a conceptual system and a cultural set of values. It is only to say that pure subjective intuition is not always our only recourse. Nor is this to say that the concepts and values of a particular culture constitute the final arbiter of fairness within the culture. There may be, and typically are, transcultural concepts and values that define a standard of fairness very different from that of a particular culture. What was fair in Nazi Germany, for example, was not fair in the eyes of the world community. Closer to home, there are court cases that constantly involve issues of fairness across subcultures with conflicting values. Here the majority culture usually gets to define fairness relative to *its* values, but these mainstream cultural values change over time and are often subject to criticism by other cultures.

What the myths of objectivism and subjectivism both miss is the way we *understand* the world through our *interactions* with it. What objectivism misses is the fact that understanding, and therefore truth, is necessarily relative to our cultural conceptual systems and that it cannot be framed in any absolute or neutral conceptual system. Objectivism also misses the fact that human conceptual systems are metaphorical in nature and involve an imaginative understanding of one kind of thing in terms of another. What subjectivism specifically misses is that our understanding, even our most imaginative understanding, is given in terms of a conceptual system that is grounded in our successful functioning in our physical and cultural environments. It also misses the fact that metaphorical understanding involves metaphorical entailment, which is an imaginative form of rationality.

26

The Myth of Objectivism in Western Philosophy and Linguistics

Our Challenge to the Myth of Objectivism

The myth of objectivism has dominated Western culture, and in particular Western philosophy, from the Pre-socratics to the present day. The view that we have access to absolute and unconditional truths about the world is the cornerstone of the Western philosophical tradition. The myth of objectivity has flourished in both the rationalist and empiricist traditions, which in this respect differ only in their accounts of how we arrive at such absolute truths. For the rationalists, only our innate capacity to reason can give us knowledge of things as they really are. For the empiricists, all our knowledge of the world arises from our sense perceptions (either directly or indirectly) and is constructed out of the elements of sensation. Kant's synthesis of rationalism and empiricism falls within the objectivist tradition also, despite his claim that there can be no knowledge whatever of things as they are in themselves. What makes Kant an objectivist is his claim that, relative to the kinds of things that all human beings can experience through their senses (his empiricist legacy), we can have universally valid knowledge and universally valid moral laws by the use of our universal reason (his rationalist legacy). The objectivist tradition in Western philosophy is preserved to this day in the descendants of the logical positivists, the Fregean tradition, the tradition of Husserl, and, in linguistics, in the neorationalism that came out of the Chomsky tradition.

Our account of metaphor goes against this tradition. We

see metaphor as essential to human understanding and as a mechanism for creating new meaning and new realities in our lives. This puts us at odds with most of the Western philosophical tradition, which has seen metaphor as an agent of subjectivism and, therefore, as subversive of the quest for absolute truth. In addition, our views on conventional metaphor—that it pervades our conceptual system and is a primary mechanism for understanding—put us at odds with the contemporary views of language, meaning, truth, and understanding that dominate recent Anglo-American analytic philosophy and go unquestioned in much of modern linguistics and other disciplines as well. The following is a representative list of these assumptions about language, meaning, truth, and understanding. Not all objectivist philosophers and linguists accept all of them, but the most influential figures seem to accept most of them.

Truth is a matter of fitting words to the world.

A theory of meaning for natural language is based on a theory of truth, independent of the way people understand and use language.

Meaning is objective and disembodied, independent of human understanding.

Sentences are abstract objects with inherent structures.

The meaning of a sentence can be obtained from the meanings of its parts and the structure of the sentence.

Communication is a matter of a speaker's transmitting a message with a fixed meaning to a hearer.

How a person understands a sentence, and what it means *to him*, is a function of the objective meaning of the sentence and what the person believes about the world and about the context in which the sentence is uttered.

Our account of conventional metaphor is inconsistent with all of these assumptions. The meaning of a sentence is given in terms of a conceptual structure. As we have seen, most of the conceptual structure of a natural language is metaphorical in nature. The conceptual structure is grounded in physical and cultural experience, as are the conventional metaphors. Meaning, therefore, is never disembodied or objective and is always grounded in the acquisition and use of a conceptual system. Moreover, truth is always given relative to a conceptual system and the metaphors that structure it. Truth is therefore not absolute or objective but is based on understanding. Thus sentences do not have inherent, objectively given meanings, and communication cannot be merely the transmission of such meanings.

It is not at all obvious why our account of these matters is so different from the standard philosophical and linguistic positions. The basic reason seems to be that all of the standard positions are based on the myth of objectivism, while our account of metaphor is inconsistent with it. Such a radical divergence from the dominant theories of such basic matters calls for explanation. How could it be possible for an account of metaphor to call into question the fundamental assumptions about truth, meaning, and understanding that have emerged from the dominant trends in the Western philosophical tradition? An answer to this requires a far more detailed account of the objectivist assumptions about language, truth, and meaning than we have given so far. It requires stating in more detail (a) what the objectivist assumptions are, (b) how they are motivated, and (c) what their implications are for a general account of language, truth, and meaning.

The point of this analysis is not merely to distinguish our views on language from the standard views but to show by example how influential the myth of objectivism is in Western culture in ways that we usually don't notice. More importantly, we want to suggest that many of the problem areas for our culture may come from a blind acceptance of the myth of objectivism and that there is another alternative short of recourse to radical subjectivity.

How Standard Theories of Meaning Are Rooted in the Myth of Objectivism

The myth of objectivism, which is the basis of the objectivist tradition, has very specific consequences for a theory of meaning. We would like to show just what these consequences are, how they arise from the myth of objectivism, and why they are untenable from an experientialist point of view. Not all objectivists hold all of the following positions, but it is common for objectivists to hold most of them in some form or other.

Meaning Is Objective

The objectivist characterizes meaning purely in terms of conditions of objective truth or falsity. On the objectivist view, the conventions of the language assign to each sentence an *objective meaning,* which determines objective truth conditions, given certain elements of context called "indexicals": who the speaker is, who his audience is, the time and place of the utterance, the objects referred to by words like "that," "this," etc. Thus, the objective meaning of a sentence does not depend on the way any given person happens to understand it or on whether he understands it at all. For example, a parrot might be trained to say "It's raining" without any understanding at all of the meaning of this in English. But the sentence has the same objective meaning whether it is said by a parrot or a person, and it will be true if it happens to be raining and false if it isn't raining. Given the objectivist account of meaning, a person understands the objective meaning of a sentence if he understands the conditions under which it would be true or false.

The objectivist assumes not only that conditions of objective truth and falsity exist but that people have access to them. This is taken as being obvious. Look around you. If there is a pencil on the floor, then the sentence "There is a pencil on the floor" is true, and, if you speak English and can perceive the pencil on the floor, you will correctly take it as being true. It is assumed that such sentences are objectively true or false and that you have access to innumerable such truths. Since people can understand the conditions under which a sentence can be objectively true, it is possible for a language to have conventions by which such objective meanings are assigned to sentences. Thus, on the objectivist view, the conventions that a language has for pairing sentences with objective meanings will depend upon speakers of the language being able to understand the sentence as having that objective meaning. Thus, when the objectivist speaks of understanding the (literal) meaning of a sentence, he is speaking of understanding what makes a sentence objectively true or false. In general, the objectivist notion of understanding is limited to understanding conditions of truth or falsity.

This is *not* what we have meant by "understanding." When we say that the objectivist views meaning as being independent of understanding, we are taking "understanding" in our sense and not his.

Meaning Is Disembodied

In the objectivist view, objective meaning is not meaning *to* anyone. Expressions in a natural language can be said to have objective meaning only if that meaning is independent of anything human beings do, either in speaking or in acting. That is, meaning must be disembodied. Frege, for example, distinguishes the "sense" (*Sinn*), the objective meaning for a sign, from the "idea," which arises

> from memories and sense impressions that I have had and acts, both internal and external, which I have performed.... The idea is subjective.... In the light of this, one need have no scruples in speaking simply of *the* sense, whereas in the case of an idea one must, strictly speaking, add to whom it belongs and at what time. [Frege, 1966, pp. 59–60]

Frege's "sense" is objective disembodied meaning. Each

linguistic expression in a language has a disembodied meaning associated with it. This is reminiscent of the CON-DUIT metaphor, where "The meaning is right there in the words."

The Fregean tradition continues to this day in the work of the disciples of Richard Montague and many others as well. In none of this work on semantics is the meaning of the sentence taken to depend in any way on the way a human being would understand it. As Montague puts it, "Like Donald Davidson, I regard the construction of a theory of truth—or rather, of the more general notion of truth under an arbitrary interpretation—as the basic goal of a serious syntax and semantics" (1974, p. 188). The important words here are "arbitrary interpretation." Montague assumed that theories of meaning and truth are purely mathematical enterprises, and his goal was to maintain an "arbitrary interpretation," untainted by anything at all having to do with human beings, especially matters of human psychology or human understanding. He intended his work to be applicable to any kind of being at all in the universe and to be free of any limitation imposed by any particular kind of being.

Fitting the Words to the World without People or Human Understanding

The objectivist tradition views semantics as the study of how linguistic expressions can fit the world directly, without the intervention of human understanding. Perhaps the clearest statement of this position is given by David Lewis:

> My proposals will also not conform to the expectations of those who, in analyzing meaning, turn immediately to the psychology and sociology of language users: to intentions, sense-experience, and mental ideas, or to social rules, conventions, and regularities. I distinguish two topics: first, the description of possible languages or grammars as abstract semantic systems whereby symbols are associated with aspects of the world; and second, the description of the psychological and

sociological facts whereby a particular one of these abstract semantic systems is the one used by a person or population. Only confusion comes of mixing these two topics. [Lewis 1972, p. 170]

Here Lewis follows the practice of Montague in trying to give an account of how language can fit the world—"how symbols are associated with aspects of the world"—that is sufficiently general and sufficiently arbitrary that it could fit any conceivable psychological or sociological facts about how people use language and how they understand it.

A Theory of Meaning Is Based on a Theory of Truth

The possibility of an account of objective truth, independent of any human understanding, makes a theory of objective meaning possible. Under the objectivist account of truth, it is possible for a sentence by itself to fit the world or not. If it does, it is true; if not, it is false. This gives rise directly to an objectivist account of meaning as based on truth. Again, David Lewis puts it most clearly: "A meaning for a sentence is something that determines the conditions under which the sentence is true or false" (1972, p. 173).

This has been generalized to give meanings for performative sentences, like orders and promises, by the technique in Lakoff (1972) and Lewis (1972). The technique uses the definition of truth in terms of "fitting the world, " which is technically defined by conditions of satisfaction in a model. Felicity conditions of speech acts are similarly defined in terms of conditions of satisfaction, or "fitting the world." When we speak of "truth" and "falsity" below, it should be understood that we are speaking in terms of conditions of satisfaction and that we are including speech acts as well as statements.

Meaning Is Independent of Use

The objectivist account of truth requires that meaning, too, be objective. If meaning is to be objective, it must exclude all subjective elements—that is, anything peculiar to a particular context, culture, or mode of understanding. As Donald Davidson puts it: "Literal meaning and truth conditions can be assigned to words and sentences apart from particular contexts of use" (1978, p. 33).

Meaning Is Compositional—The Building-Block Theory

According to the myth of objectivism, the world is made up of objects; they have well-defined inherent properties, independent of any being who experiences them, and there are fixed relations holding among them at any given point in time. These aspects of the myth of objectivism give rise to a building-block theory of meaning. If the world is made up of well-defined objects, we can give them names in a language. If the objects have well-defined inherent properties, we can have a language with one-place predicates corresponding to each of those properties. And if the objects stand in fixed relations to one another (at least at any given instant), we can have a language with many-place predicates corresponding to each relation.

Assuming that the world is this way and that we have such a language, we can, using the syntax of this language, construct sentences that can correspond directly to any situation in the world. The meaning of the whole sentence will be its truth conditions, that is, the conditions under which the sentence can be fitted to some situation. The meaning of the whole sentence will depend entirely on the meanings of its parts and how they fit together. The meanings of the parts will specify what names can pick out what objects and what predicates can pick out what properties and relations.

Objectivist theories of meaning are all compositional in nature—that is, they are all building-block theories—and they have to be. The reason is that, for the objectivist, the world *is* made up of building blocks: definable objects and clearly delineated inherent properties and relations.

Moreover, every sentence of the language must contain all of the necessary building blocks so that, together with the syntax, nothing more is needed to provide the truth conditions of the sentence. The "something more" that is ruled out is any kind of human understanding.

Objectivism Permits Ontological Relativity without Human Understanding

The logical positivists (e.g., Carnap) attempted to carry out an objectivist program by trying to construct a universally applicable formal (logical) language that had all of the building-block properties mentioned above and all of the other characteristics we have discussed so far. Richard Montague (1974) claimed to have provided a "universal grammar" that would map natural languages onto such a universally applicable formal language.

Quine, reacting to such universalist claims, argued that each language has its own ontology built into it, and what counts as an object, property, or relation may vary from language to language. This position is known as the "ontological relativity" thesis.

It is possible to maintain an ontological relativity thesis within the confines of the objectivist program without any recourse to human understanding or cultural difference. Such a relativistic position gives up on the possibility of constructing a single universally applicable logical language into which all natural languages can be translated adequately. It claims instead that each natural language carves up what is in the world in different ways—always picking out objects that are really there and properties and relations that are really there. But since different languages *may* have different ontologies built in, there is no guarantee that any two languages will, in general, be commensurable.

The relativistic version of the objectivist account of meaning thus claims that meaning and truth conditions are objectively given, not in universal terms, but only relative to a given language. This relativistic objectivism still holds to the myth of objectivism in claiming that truth is objective and that there are objects in the world with inherent properties. But, according to relativistic objectivism, truths expressible in one language may not be translatable into another, since each language may carve up the world in different ways. But whatever entities the language picks out exist in the world objectively as entities. Truth and meaning are still objective in this account (though relative to a given language), and human understanding is still ruled out as irrelevant to meaning and truth.

Linguistic Expressions Are Objects: The Premise of Objectivist Linguistics

According to the myth of objectivism, objects have properties in and of themselves and they stand in relationships to one another independently of any being who understands them. When words and sentences are written down, they can be readily looked upon as objects. This has been the premise of objectivist linguistics from its origins in antiquity to the present: Linguistic expressions are objects that have properties in and of themselves and stand in fixed relationships to one another, independently of any person who speaks them or understands them. As objects, they have parts—they are made up of building blocks: words are made up of roots, prefixes, suffixes, infixes; sentences are made up of words and phrases; discourses are made up of sentences. Within a language, the parts can stand in various relationships to one another, depending upon their building-block structure and their inherent properties. The study of the building-block structure, the inherent properties of the parts, and the relationships among them has traditionally been called *grammar*.

Objectivist linguistics sees itself as the only *scientific* approach to linguistics. The objects must be capable of being analyzed in and of themselves, independently of contexts or the way people understand them. As in objectivist philosophy, there are both empiricist and rationalist traditions in linguistics. The empiricist tradition, represented by the latter-day American structuralism of Bloomfield, Harris, and their followers, took texts as the only objects of scientific study. The rationalist tradition, represented by European structuralists such as Jakobson and American figures like Sapir, Whorf, and Chomsky, viewed language as having mental reality, with linguistic expressions as mentally real objects.

Grammar Is Independent of Meaning and Understanding

We have just seen how the myth of objectivism gives rise to a view of language in which linguistic expressions are objects with inherent properties, a building-block structure, and fixed relationships among the objects. According to the myth of objectivism, the linguistic objects that exist—and their building-block structure, their properties, and their relations—are independent of the way people understand them. It follows from this view of linguistic expressions as objects that grammar can be studied independently of meaning or human understanding.

This tradition is epitomized by the linguistics of Noam Chomsky, who has steadfastly maintained that grammar is a matter of pure form, independent of meaning or human understanding. Any aspect of language that involves human understanding is for Chomsky by definition outside the study of grammar in this sense. Chomsky's use of the term "competence" as opposed to "performance" is an attempt to define certain aspects of language as the only legitimate objects of what he considers scientific linguistics—that is, what we have called objectivist linguistics in the rationalist mode, including only matters of pure form and excluding all matters of human understanding and language use. Though Chomsky sees linguistics as a branch of psychology, it is for him an *independent* branch, one that is in no way dependent on the way people actually understand language.

The Objectivist Theory of Communication: A Version of the CONDUIT Metaphor

Within objectivist linguistics and philosophy, meanings and linguistic expressions are independently existing objects. Such a view gives rise to a theory of communication that fits the CONDUIT metaphor very closely:

Meanings are objects.
Linguistic expressions are objects.
Linguistic expressions have meanings (in them).
In communication, a speaker sends a fixed meaning to a hearer via the linguistic expression associated with that meaning.

On this account it is possible to objectively say what you mean, and communication failures are matters of subjective errors: since the meanings are objectively right there in the words, either you didn't use the right words to say what you meant or you were misunderstood.

What an Objectivist Account of Understanding Would Be Like

We have already given an account of what the objectivist means by understanding the literal objective meaning of a sentence, namely, understanding the conditions under which a sentence would be objectively true or false. Objectivists recognize, however, that a person may understand a sentence in a given context as meaning something other than its literal objective meaning. This other meaning is usually called the "speaker's meaning" or the "utterer's meaning," and objectivists typically recognize that any full account of understanding will have to account for these cases, too (see Grice 1957).

Take, for example, the sentence "He's a real genius," uttered in a context where sarcasm is clearly indicated. On the objectivist account, there is an *objective meaning* of the sentence "He's a real genius," namely, that he has great intellectual powers. But in uttering the sentence sarcastically, the speaker intends to convey the opposite meaning, namely, that he's an utter idiot. The *speaker's meaning* here is the opposite of the objective meaning of the sentence.

This account of speaker's meaning could be represented, in the appropriate sarcastic context, as follows:

(A) In uttering a sentence S (S = "He's a real genius"), which has the objective meaning M (M = he has great intellectual powers), the speaker intends to convey to the hearer objective meaning M' (M' = he's a real idiot).

This is how meaning *to* someone might be accounted for in an objectivist framework. Sentence (A) is something that could be objectively true or false in a given context. If (A) is true, then the sentence S ("He's a real genius") can mean *he's a real idiot* to both the speaker and the hearer if the hearer recognizes the speaker's intentions.

This technique, which originated with the speech-act theorists, has been adapted to the objectivist tradition as a way of getting meaning *to* someone out of the objective meaning of the sentence, that is, out of its conditions for objective truth or falsity. The technical trick here involves using two objective meanings, M and M', together with sentence (A), which also has an objective meaning, in such a way as to get an account of speaker's meaning and hearer's meaning, that is, meaning *to* someone. This, of course, involves recognizing a speaker's intentions as being objectively real, which some objectivists might deny.

The example we have given is one of sarcasm, where M and M' have opposite meanings, that is, opposite truth conditions. Speaking literally would be a case where M = M'. The objectivist program sees this as a general technique for accounting for all cases of meaning *to* a person, especially where a speaker says one thing and means something else: exaggeration, understatement, hints, irony, and all figurative language—in particular, metaphor. Carrying out the program would involve formulating general principles that would answer the following question:

Given sentence S and its *literal* objective meaning M, and given the relevant knowledge of the context, what specific principles allow us to predict what the *speaker's meaning* M' will be in this context?

In particular, this applies in the case of metaphor. For example, "This theory is made of cheap stucco" would, on the objectivist account, have a literal objective meaning (M) which is false, namely, *this theory is made of inexpensive mortar*. The literal objective meaning is false because theories are not the kind of thing that can be made up of mortar at all. However, "This theory is made of cheap stucco" could have an intended speaker's meaning (M') which might be true, namely, *this theory is weak*. In this case, the problem would be to give general principles of interpretation by which a hearer could move from the sentence S ("This theory is made of cheap stucco") to the intended speaker's meaning M (*this theory is weak*) via the objective meaning M (*this theory is made of inexpensive mortar*).

The objectivist sees all metaphors as cases of indirect meaning, where M ≠ M'. All sentences containing metaphors have objective meanings that are, in the typical case, either blatantly false (e.g., "The theory is made of cheap stucco") or blatantly true (e.g., "Mussolini was an animal"). Understanding a sentence (e.g., "The theory is made of cheap stucco") as metaphorical always involves understanding it indirectly as conveying an objective meaning M' (*the theory is weak*) which is different from the literal objective meaning M (*the theory is made of inexpensive mortar*).

The objectivist account of understanding is thus always based on its account of objective truth. It includes two kinds of understanding, direct and indirect. Direct understanding is understanding a literal objective meaning of a sentence in terms of the conditions under which it can be objectively true. Indirect understanding involves figuring out when the speaker is using one sentence to convey an indirect meaning, where the conveyed meaning can be *understood directly* in terms of objective truth conditions.

There are four automatic consequences of the objectivist account of metaphor:

> *By definition, there can be no such thing as a metaphorical concept or metaphorical meaning.* Meanings are objective and specify conditions of objective truth. They are by definition ways of characterizing the world as it is or might be. Conditions of objective truth simply do not provide ways of viewing one thing in terms of another. Hence, objective meanings cannot be metaphorical.

> *Since metaphor cannot be a matter of meaning, it can only be a matter of language.* A metaphor, on the objectivist view, can at best give us an indirect way of *talking* about some objective meaning M' by using the language that would be used literally to talk about some other objective meaning M, which is usually false in a blatant way.

> *Again by definition, there can be no such thing as literal (conventional) metaphor.* A sentence is used literally when M' = M, that is, when the speaker's meaning is the objective meaning. Metaphors can only arise when M' ≠ M. Thus, according to the objectivist definition, a literal metaphor is a contradiction in terms, and literal language cannot be metaphorical.

> *Metaphor can contribute to understanding only by making us see objective similarities, that is, similarities between the objective meanings M and M'.* These similarities must be based on shared *inherent properties* of objects—properties that the objects really have, in and of themselves.

Thus, the objectivist account of meaning is completely at odds with everything we have claimed in this book. This view of meaning and of metaphor has been with us since the time of the Greeks. It fits the CONDUIT metaphor ("The meaning is right there in the words") and it fits the myth of objectivism.

27

How Metaphor Reveals the Limitations of the Myth of Objectivism

The heart of the objectivist tradition in philosophy comes directly out of the myth of objectivism: the world is made up of distinct objects, with inherent properties and fixed relations among them at any instant. We argue, on the basis of linguistic evidence (especially metaphor), that the objectivist philosophy fails to account for the way we understand our experience, our thoughts, and our language. An adequate account, we argue, requires

—viewing objects only as entities relative to our interactions with the world and our projections on it

—viewing properties as interactional rather than inherent

—viewing categories as experiential gestalts defined via prototype instead of viewing them as rigidly fixed and defined via set theory

We view issues having to do with meaning in natural language and with the way people understand both their language and their experiences as empirical issues rather than matters of a priori philosophical assumptions and argumentation. We have selected metaphor and the way we understand it from among the possible domains of evidence that could bear on these issues. We have focused on metaphor for the following four reasons:

In the objectivist tradition, metaphor is of marginal interest at best, and it is excluded altogether from the study of semantics (objective meaning). It is seen as only marginally relevant to an account of truth.

Yet we have found that metaphor is pervasive, not

merely in our language but in our conceptual system. It seems inconceivable to us that any phenomenon so fundamental to our conceptual system could not be central to an account of truth and meaning.

We observed that metaphor is one of the most basic mechanisms we have for understanding our experience. This did not jibe with the objectivist view that metaphor is of only peripheral interest in an account of meaning and truth and that it plays at best a marginal role in understanding.

We found that metaphor could create new meaning, create similarities, and thereby define a new reality. Such a view has no place in the standard objectivist picture of the world.

The Objectivist Account of Conventional Metaphor

Many of the facts that we have discussed have long been known in the objectivist tradition, but they have been given an entirely different interpretation from ours.

The conventional metaphorical concepts we take as structuring our everyday conceptual system are taken by the objectivists to be nonexistent. Metaphors, for them, are matters of mere language; there are no such things as metaphorical concepts.

Words and expressions that we have taken as instances of metaphorical concepts (e.g., *digest* in "I can't digest all those facts") would be taken by objectivists as not being instances of live metaphor at all. For them the word *digest* would have two different and distinct literal (objective) meanings—*digest*$_1$ for food and *digest*$_2$ for ideas. On this account, there would be two words *digest* which are homonyms, like the two words *bank* (*bank of a river* and *bank where you put your money*).

An objectivist might grant that *digest an idea* was *once* a metaphor, but he would claim that it is no longer metaphorical. For him it is a "dead metaphor," one that has become conventionalized and has its own literal meaning. This is to say that there are two homonymous words *digest*.

The objectivist would probably grant that *digest*₁ and *digest*₂ have similar meanings and that the similarity is the basis for the original metaphor. This, he would say, explains why the same word is used to express two different meanings; it was once a metaphor, it became a conventionalized part of the language; it died and became frozen, taking its old metaphorical meaning as a new literal meaning.

The objectivist would observe that the similarities upon which the dead metaphor was based can in many cases still be perceived today.

According to the objectivist account of metaphor, the original metaphor was a matter of use and speaker's meaning, not literal objective meaning. It would have to have arisen by the general speaker's meaning formula applied to this case (where *digest* referred only to food):

> In uttering a sentence S (S = "I couldn't digest his ideas") with literal objective meaning M (M = I couldn't transform his ideas, by chemical and muscular action in the alimentary canal, into a form my body could absorb), the speaker intends to convey to the hearer the speaker's meaning M' (M' = I couldn't transform his ideas, by mental action, into a form my mind could absorb).

Two things have to be true in order for this objectivist account to hold. First, the intended speaker's meaning M', referring to ideas, must be an objectively given meaning, having objective truth conditions. In other words, the following must be *objectively* true of the mind and ideas by virtue of their inherent properties:

> Ideas must, by virtue of their inherent properties, be the kind of thing that can have a form, be transformed, and be absorbed into the mind.
>
> The mind must, by virtue of its inherent properties, be the kind

of thing that can perform mental actions, transform ideas, and absorb them into itself.

Second, the metaphor must have been originally based on preexisting similarities between M and M'. That is, the mind and the alimentary canal must have *inherent* properties in common, just as ideas and food must have *inherent* properties in common.

To summarize: the dead-metaphor account of *digest* would claim the following:

The word *digest* originally referred to a food concept.

By a "live" metaphor, the word *digest* was transferred to a preexisting objective meaning in the realm of ideas, on the basis of preexisting objective similarities between food and ideas.

Eventually the metaphor "died," and the metaphorical use of *digest un idea* became conventional. *Digest* thus obtained a second literal objective meaning, the one occurring in M'. This is seen, on the objectivist account, as a typical way of providing words for preexisting meanings that lack words to express them. All such cases would be considered homonyms.

In general, an objectivist would have to treat *all* of our conventional-metaphor data according to either the homonymy position (typically the weak version) or the abstraction position. Both of these positions depend on the existence of preexisting similarities based on inherent properties.

What's Wrong with the Objectivist Account

As we have just seen, the objectivist account of conventional metaphor requires either an abstraction view or a homonymy view. Moreover, the objectivist account of both conventional and nonconventional metaphor is based on preexisting inherent similarities. We have already presented detailed arguments against all of these positions. These arguments take on a special importance now. *They show not only that the objectivist view of metaphor is in-*

adequate but that the entire objectivist program is based on erroneous assumptions. To see just where the objectivist account of metaphor is inadequate, let us recall the relevant parts of our arguments against the abstraction, homonymy, and similarity views as they pertain to the objectivist account of *conventional* metaphor.

The Similarity Position

We saw in our discussion of the IDEAS ARE FOOD metaphor that, although the metaphor was based on similarities, the similarities themselves were not inherent but were based on other metaphors—in particular, THE MIND IS A CONTAINER, IDEAS ARE OBJECTS, and the CONDUIT metaphors. The view that IDEAS ARE OBJECTS is a projection of entity status upon mental phenomena via an ontological metaphor. The view that THE MIND IS A CONTAINER is a projection of entity status with in-out orientation onto our cognitive faculty. These are not *inherent objective properties* of ideas and of the mind. They are *interactional properties,* and they reflect the way in which we *conceive* of mental phenomena by virtue of metaphor.

The same holds in the case of our concepts TIME and LOVE. We understand sentences like "The time for action has arrived" and "We need to budget our time" in terms of the TIME IS A MOVING OBJECT and TIME IS MONEY metaphors, respectively. But on the objectivist account there would be no such metaphors. *Arrive* and *budget* in these sentences would be dead metaphors, that is, homonyms, deriving historically from once-live metaphors. These once-live metaphors would have to have been based on inherent similarities between time and moving objects, on the one hand, and time and money, on the other. But, as we have seen, such similarities are not inherent; they are themselves created via ontological metaphors.

It is even more difficult to make a case for an inherent-similarity analysis for expressions involving the concept LOVE, such as "This relationship isn't going anywhere,"

"There was a magnetism between us," and "This relationship is dying." The concept LOVE is not inherently
well defined. Our culture gives us conventional ways of
viewing love experiences via conventional metaphors, such
as LOVE IS A JOURNEY, LOVE IS A PHYSICAL FORCE, etc., and
our language reflects these. But according to the objectivist
account (based either on dead metaphor, weak homonymy,
or abstraction), the concept LOVE must be sufficiently well
defined in terms of inherent properties to bear inherent
similarities to journeys, electromagnetic and gravitational
phenomena, sick people, etc. Here the objectivist must not
only bear the burden of claiming that love has inherent
properties similar to the inherent properties of journeys,
electromagnetic phenomena, and sick people; he must also
claim that love is sufficiently clearly defined in terms of
these inherent properties so that those similarities will
exist.

In summary, the usual objectivist accounts of these
phenomena (dead metaphor, homonymy with similarities,
or abstraction) all depend on preexisting similarities based
on inherent properties. In general, similarities do exist, but
they cannot be based on *inherent* properties. The
similarities arise as a *result* of conceptual metaphors and
thus must be considered similarities of *interactional,* rather
than inherent, properties. But the admission of interactional
properties is inconsistent with the basic premise of objectivist philosophy. It amounts to giving up the myth of objectivism.

The Objectivist Default: "It's Not Our Job"

The only remaining alternative for the objectivist is to give
up any account of any relationship between the FOOD and
IDEA senses of *digest* in terms of similarity (including denial
that there was ever a metaphor at all) and to turn to the
strong homonymy position. According to this view there is
one word *digest* with two entirely different and unrelated
meanings—as different as the two meanings of *punt (a kick*

in football and *an open, flat-bottomed boat with square ends*). As we have seen (in chapter 18), the strong homonymy position cannot account for:

Internal systematicity
External systematicity
Extensions of the used portion of the metaphor
The use of concrete experience to structure abstract experience
The similarities that we, in fact, see between the two senses of
    digest, based on metaphorically conceptualizing ideas in
    terms of food.

Of course, an objectivist philosopher or linguist could grant that he cannot adequately account for such systematicities, similarities, and ways of understanding the less concrete in terms of the more concrete. This might not disturb him in the slightest. After all, he could claim, accounting for such things is not his job. Such things are matters for the psychologist, the neurophysiologist, the philologist, or someone else. This would be in the tradition of Frege's separating off "sense" from "ideas" and Lewis's separating off "abstract semantic systems" from "psychological and sociological facts." The homonymy view, they could claim, is adequate for their proper purposes as objectivists, namely, to provide objective truth conditions for linguistic expressions and to give an account of literal objective meaning in terms of them. This, they assume, could be done independently for the two senses of *digest* without having to account for systematicity, similarity, understanding, etc. Relative to this conception of their job, conventional metaphorical uses of *digest* involve merely homonyms and not metaphors at all, dead or alive. The only metaphors they recognize are nonconventional metaphors (e.g., "Your ideas are made of cheap stucco" or "Love is a collaborative work of art"). Since these, they would claim, are matters of speaker's meaning, not the literal objective meaning of a sentence, issues of truth and meaning arising from them are to be handled by the account of speaker's meaning given above.

In summary, the only internally consistent objectivist view of conventional metaphor would be that the issues we have been primarily concerned with—the properties of conventional metaphors and the way we use them in understanding—are simply outside their purview. They would insist that they are not responsible for such matters and that no facts of this sort concerning conventional metaphor could possibly have any bearing on the objectivist program or on what they, as objectivists, believe.

Such objectivists might even grant that our investigations of metaphor correctly show that *interactional properties* and *experiential gestalts* are, in fact, necessary to account for how human beings understand their experience via metaphor. But even granting this, they could still continue to ignore everything we have done on the following grounds: they could say simply that experientialists are merely concerned with how human beings happen to understand reality, given all of their limitations, whereas the objectivist is concerned not with how people *understand* something as being true but rather with what it means for something to *actually be true*.

This objectivist response perfectly highlights the fundamental difference between objectivism and experientialism. Such an objectivist reply boils down to a reaffirmation of their fundamental concern with "absolute truth" and "objective meaning," entirely independent of anything having to do with human functioning or understanding. Against this, we have been maintaining that there is no reason to believe that there is any absolute truth or objective meaning. Instead, we maintain that it is possible to give an account of truth and meaning only relative to the way people function in the world and understand it. We are simply in a different philosophical universe from such objectivists.

The Irrelevance of Objectivist Philosophy to Human Concerns

We are in the same philosophical universe as, and have real disagreements with, those objectivists who think that there can be an adequate objectivist account of human understanding, of our conceptual system and our natural language. We have argued in detail that conventional metaphor is pervasive in human language and the human conceptual system and that it is a primary vehicle for understanding. We have argued that an adequate account of understanding requires interactional properties and experiential gestalts. Since all objectivist accounts require inherent properties and most of them require a set-theoretical account of categorization, they fail to give an adequate account of how human beings conceptualize the world.

Objectivist Models Outside of Objectivist Philosophy

Classical mathematics comprises an objectivist universe. It has entities that are clearly distinguished from one another, e.g., numbers. Mathematical entities have inherent properties, e.g., *three* is odd. And there are fixed relationships among those entities, e.g., *nine* is the square of *three*. Mathematical logic was developed as part of the enterprise of providing foundations for classical mathematics. Formal semantics also developed out of that enterprise. The models used in formal semantics are examples of what we will call ''objectivist models''—models appropriate to universes of discourse where there are distinct entities which have inherent properties and where there are fixed relationships among the entities.

But the real world is not an objectivist universe, especially those aspects of the real world having to do with human beings: human experience, human institutions, human language, the human conceptual system. What it means to be a hard-core objectivist is to claim that there is an objectivist model that fits the world as it really is. We have just argued that objectivist philosophy is empirically incorrect in that it makes false predictions about language, truth, understanding, and the human conceptual system. On the basis of this we have claimed that objectivist philosophy provides an inadequate basis for the human sciences.

Nonetheless, a lot of remarkably insightful mathematicians, logicians, linguists, psychologists, and computer scientists have designed objectivist models for use in the human sciences. Are we claiming that all of this work is worthless and that objectivist models have no place at all in the human sciences?

We are claiming no such thing. We believe that objectivist models as mathematical entities do not necessarily have to be tied to objectivist philosophy. One can believe that objectivist models can have a function—even an important function—in the human sciences without adopting the objectivist premise that there is an objectivist model that completely and accurately fits the world as it really is. But if we reject this premise, what role is left for objectivist models?

Before we can answer this question, we need to look at some of the properties of ontological and structural metaphors:

Ontological metaphors are among the most basic devices we have for comprehending our experience. Each structural metaphor has a consistent set of ontological metaphors as subparts. To use a set of ontological metaphors to comprehend a given situation is to impose an entity structure upon that situation. For example, LOVE IS A JOURNEY imposes on LOVE an entity structure including a beginning, a destination, a path, the distance you are along the path, and so on.

Each individual structural metaphor is internally consistent and imposes a consistent structure on the concept it structures. For example, the ARGUMENT IS WAR metaphor imposes an internally consistent WAR structure on the concept ARGUMENT. When we understand love only in terms of the LOVE IS A JOURNEY metaphor, we are imposing an internally consistent JOURNEY structure on the concept LOVE.

Although different metaphors for the same concept are not in general consistent with each other, it is possible to find sets of metaphors that *are* consistent with each other. Let us call these *consistent sets of metaphors*.

Because each individual metaphor is internally consistent, each consistent set of metaphors allows us to comprehend a situation in terms of a well-defined entity structure with consistent relations between the entities.

The way that a consistent set of metaphors imposes an entity structure with a set of relations between the entities can be represented by an objectivist model. In the model, the entities are those imposed by the ontological metaphors, and the relations between the entities are those given by the internal structures of the structural metaphors.

To summarize: Trying to structure a situation in terms of such a consistent set of metaphors is in part like trying to structure that situation in terms of an objectivist model. What is left out are the experiential bases of the metaphors and what the metaphors hide.

The natural question to ask, then, is whether people actually think and act in terms of consistent sets of metaphors. A special case where they do is in the formulation of scientific theories, say, in biology, psychology, or linguistics. Formal scientific theories are attempts to consistently extend a set of ontological and structural metaphors. But in addition to scientific theorizing, we feel that people do try to think and act in terms of consistent sets of metaphors in a wide variety of situations. These are cases where people might be viewed as trying to apply objectivist models to their experience.

There is an excellent reason for people to try to view a life situation in terms of an objectivist model, that is, in terms of a consistent set of metaphors. The reason is, simply, that if we can do this, we can draw inferences about the situation that will not conflict with one another. That is, we will be able to infer nonconflicting expectations and suggestions for behavior. And it is comforting—extremely comforting—to have a consistent view of the world, a clear set of expectations and no conflicts about what you should do. Objectivist models have a real appeal—and for the most human of reasons.

We do not wish to belittle this appeal. It is the same as the appeal of finding coherence in your life or in some range of life experiences. Having a basis for expectation and action is important for survival. But it is one thing to impose a single objectivist model in some restricted situations and to function in terms of that model—perhaps successfully; it is another to conclude that the model is an accurate reflection of reality. There is a good reason why our conceptual systems have inconsistent metaphors for a single concept. The reason is that there is no one metaphor that will do. Each one gives a certain comprehension of one aspect of the concept and hides others. To operate only in terms of a consistent set of metaphors is to hide many aspects of reality. Successful functioning in our daily lives seems to require a constant shifting of metaphors. The use of many metaphors that are inconsistent with one another seems necessary for us if we are to comprehend the details of our daily existence.

One obvious utility for the study of formal objectivist models in the human sciences is that they can allow us to understand, *in part,* the ability to reason and function in terms of a consistent set of metaphors. This is a common activity and an important one to understand. It can also allow us to see what can be wrong with imposing a requirement of consistency—to see that any consistent set of metaphors will most likely hide indefinitely many aspects of reality—aspects that can be highlighted only by other metaphors that are inconsistent with it.

One obvious limitation of formal models is that, so far as we can imagine, they provide no means for including the experiential basis for a metaphor and therefore provide no way of accounting for the way in which metaphorical concepts permit us to comprehend our experience. There is a corollary of this that has to do with the issue of whether a computer could ever understand things the way people do. The answer we give is no—simply because understanding requires experience, and computers don't have bodies and don't have human experiences.

However, the study of computational models might nevertheless tell us a great deal about human intellectual capacities, especially in the areas where people reason and function partly in terms of objectivist models. Moreover, current formal techniques in computer science show promise of providing representations of *inconsistent* sets of metaphors. This could conceivably lead to insights about the way that people reason and function in terms of coherent, but inconsistent, metaphorical concepts. The limits of formal study seem to be in the area of the experiential bases of our conceptual system.

Summary

Our general conclusion is that the objectivist program is unable to give a satisfactory account of human understanding and of any issues requiring such an account. Among these issues are:

—the human conceptual system and the nature of human rationality
—human language and communication
—the human sciences, especially psychology, anthropology, sociology, and linguistics
—moral and aesthetic value
—scientific understanding, via the human conceptual system
—any way in which the foundations of mathematics have a basis in human understanding

The basic elements of an experientialist account of understanding—interactional properties, experiential gestalts, and metaphorical concepts—seem to be necessary for any adequate treatment of these human issues.

28

Some Inadequacies of the Myth of Subjectivism

In Western culture, the chief alternative to objectivism has traditionally been taken to be subjectivism. We have argued that the myth of objectivism is inadequate to account for human understanding, human language, human values, human social and cultural institutions, and everything dealt with by the human sciences. Thus, according to the dichotomy that our culture would foist upon us, we would be left only with a radical subjectivity, which denies the possibility of any scientific "lawlike" account of human realities.

But we have claimed that subjectivism is not the only alternative to objectivism, and we have been offering a third choice: the experientialist myth, which we see as making possible an adequate philosophical and methodological basis for the human sciences. We have already distinguished this alternative from the objectivist program, and it is equally important to distinguish it from a subjectivist program.

Let us consider briefly some subjectivist positions on how people understand their experience and their language. These flow mainly from the Romantic tradition and are to be found in contemporary interpretations (probably *mis*-interpretations) of recent Continental philosophy, especially the traditions of phenomenology and existentialism. Such subjectivist interpretations are largely popularizations that pick and choose elements of antiobjectivist Continental philosophy, often ignoring what makes certain trends in Continental thought serious attempts to provide a basis for the human sciences. These subjectivist positions, listed

below, might be characterized jointly as "café phenomenology." They include:

Meaning is private: Meaning is always a matter of what is
meaningful and significant *to* a person. What an individual finds
significant and what it means *to* him are matters of intuition,
imagination, feeling, and individual experience. What something means to one individual can never be fully known or
communicated to anyone else.

Experience is purely holistic: There is no natural structuring to
our experience. Any structure that we or others place on our
experience is completely artificial.

Meanings have no natural structure: Meaning to an individual
is a matter of his private feelings, experiences, intuitions, and
values. These are purely holistic; they have no natural structure. Thus, meanings have no natural structure.

Context is unstructured: The context needed for understanding
an utterance—the physical, cultural, personal, and interpersonal context—has no natural structure.

Meaning cannot be naturally or adequately represented: This
is a consequence of the facts that meanings have no natural
structure, that they can never be fully known or communicated
to another person, and that the context needed to understand
them is unstructured.

These subjectivist positions all hinge on one basic assumption, namely, that experience has no natural structure
and that, therefore, there can be no natural external constraints upon meaning and truth. Our reply follows directly
from our account of how our conceptual system is
grounded. We have argued that our experience is structured
holistically in terms of experiential gestalts. These gestalts
have structure that is not arbitrary. Instead, the dimensions
that characterize the structure of the gestalts emerge naturally from our experience.

This is not to deny the possibility that what something
means to me may be based on *kinds* of experiences that I
have had and you have not had and that, therefore, I will

not be able to *fully* and *adequately* communicate that meaning to you. However, metaphor provides a way of *partially* communicating unshared experiences, and it is the natural structure of our experience that makes this possible.

29

The Experientialist Alternative:
Giving New Meaning to the Old
Myths

The fact that the myths of subjectivism and objectivism have stood for so long in Western culture indicates that each serves some important function. Each myth is motivated by real and reasonable concerns, and each has some grounding in our cultural experience.

What Experientialism Preserves of the Concerns That Motivate Objectivism

The fundamental concern of the myth of objectivism is the world external to the individual. The myth rightly emphasizes the fact that there are real things, existing independently of us, which constrain both how we interact with them and how we comprehend them. Objectivism's focus on truth and factual knowledge is based on the importance of such knowledge for successful functioning in our physical and cultural environment. The myth is also motivated by a concern for fairness and impartiality in cases where that matters and can be achieved in some reasonable fashion.

The experientialist myth, as we have been sketching it, shares all these concerns. Experientialism departs from objectivism, however, on two fundamental issues:

Is there an absolute truth?

Is absolute truth necessary to meet the above concerns—the concern with knowledge that allows us to function successfully and the concern with fairness and impartiality?

Experientialism answers no to both questions. Truth is

always relative to understanding, which is based on a nonuniversal conceptual system. But this does not preclude satisfying the legitimate concerns about knowledge and impartiality that have motivated the myth of objectivism for centuries. Objectivity is still possible, but it takes on a new meaning. Objectivity still involves rising above individual bias, whether in matters of knowledge or value. But where objectivity is reasonable, it does not require an absolute, universally valid point of view. Being objective is always relative to a conceptual system and a set of cultural values. Reasonable objectivity may be impossible when there are conflicting conceptual systems or conflicting cultural values, and it is important to be able to admit this and to recognize when it occurs.

According to the experientialist myth, scientific knowledge is still possible. But giving up the claim to absolute truth could make scientific practice more responsible, since there would be a general awareness that a scientific theory may hide as much as it highlights. A general realization that science does not yield absolute truth would no doubt change the power and prestige of the scientific community as well as the funding practices of the federal government. The result would be a more reasonable assessment of what scientific knowledge is and what its limitations are.

What Experientialism Preserves of the Concerns That Motivate Subjectivism

What legitimately motivates subjectivism is the awareness that meaning is always meaning *to* a person. What's meaningful to me is a matter of what has significance for me. And what is significant for me will not depend on my rational knowledge alone but on my past experiences, values, feelings, and intuitive insights. Meaning is not cut and dried; it is a matter of imagination and a matter of constructing coherence. The objectivist emphasis on achieving a universally valid point of view misses what is important, insightful, and coherent for the individual.

The experientialist myth agrees that understanding does involve all of these elements. Its emphasis on interaction and interactional properties shows how meaning always is meaning *to* a person. And its emphasis on the construction of coherence via experiential gestalts provides an account of what it means for something to be significant to an individual. Moreover, it gives an account of how understanding uses the primary resources of the imagination via metaphor and how it is possible to give experience new meaning and to create new realities.

Where experientialism diverges from subjectivism is in its rejection of the Romantic idea that imaginative understanding is completely unconstrained.

In summary, we see the experientialist myth as capable of satisfying the real and reasonable concerns that have motivated the myths of both subjectivism and objectivism but without either the objectivist obsession with absolute truth or the subjectivist insistence that imagination is totally unrestricted.

30

Understanding

We see a single human motivation behind the myths of both objectivism and subjectivism, namely, a concern for understanding. The myth of objectivism reflects the human need to understand the *external* world in order to be able to function successfully in it. The myth of subjectivism is focused on *internal* aspects of understanding—what the individual finds meaningful and what makes his life worth living. The experientialist myth suggests that these are not opposing concerns. It offers a perspective from which both concerns can be met at once.

The old myths share a common perspective: man as separate from his environment. Within the myth of objectivism, the concern for truth grows out of a concern for successful functioning. Given a view of man as separate from his environment, successful functioning is conceived of as *mastery over* the environment. Hence, the objectivist metaphors KNOWLEDGE IS POWER and SCIENCE PROVIDES CONTROL OVER NATURE.

The principal theme of the myth of subjectivism is the attempt to overcome the alienation that results from viewing man as separate from his environment and from other men. This involves an embracing of the self—of individuality and reliance upon personal feelings, intuition, and values. The Romanticist version involves reveling in the senses and the feelings and attempting to gain union with nature through passive appreciation of it.

The experientialist myth takes the perspective of man as part of his environment, not as separate from it. It focuses

229

on constant interaction with the physical environment and with other people. It views this interaction with the environment as involving mutual change. You cannot function within the environment without changing it or being changed by it.

Within the experientialist myth, understanding emerges from interaction, from constant negotiation with the environment and other people. It emerges in the following way: the nature of our bodies and our physical and cultural environment imposes a structure on our experience, in terms of natural dimensions of the sort we have discussed. Recurrent experience leads to the formation of categories, which are experiential gestalts with those natural dimensions. Such gestalts define coherence in our experience. We understand our experience directly when we see it as being structured coherently in terms of gestalts that have emerged directly from interaction with and in our environment. We understand experience metaphorically when we use a gestalt from one domain of experience to structure experience in another domain.

From the experientialist perspective, truth depends on understanding, which emerges from functioning in the world. It is through such understanding that the experientialist alternative meets the objectivist's need for an account of truth. It is through the coherent structuring of experience that the experientialist alternative satisfies the subjectivist's need for personal meaning and significance.

But experientialism provides more than just a synthesis that meets the motivating concerns of objectivism and subjectivism. The experientialist account of understanding provides a richer perspective on some of the most important areas of experience in our everyday lives:

Interpersonal communication and mutual understanding
Self-understanding
Ritual
Aesthetic experience
Politics

We feel that objectivism and subjectivism both provide impoverished views of all of these areas because each misses the motivating concerns of the other. What they both miss in all of these areas is an interactionally based and creative understanding. Let us turn to an experientialist account of the nature of understanding in each of these areas.

Interpersonal Communication and Mutual Understanding

When people who are talking don't share the same culture, knowledge, values, and assumptions, mutual understanding can be especially difficult. Such understanding *is* possible through the negotiation of meaning. To negotiate meaning with someone, you have to become aware of and respect both the differences in your backgrounds and when these differences are important. You need enough diversity of cultural and personal experience to be aware that divergent world views exist and what they might be like. You also need patience, a certain flexibility in world view, and a generous tolerance for mistakes, as well as a talent for finding the right metaphor to communicate the relevant parts of unshared experiences or to highlight the shared experiences while deemphasizing the others. Metaphorical imagination is a crucial skill in creating rapport and in communicating the nature of unshared experience. This skill consists, in large measure, of the ability to bend your world view and adjust the way you categorize your experience. Problems of mutual understanding are not exotic; they arise in all extended conversations where understanding is important.

When it really counts, meaning is almost never communicated according to the CONDUIT metaphor, that is, where one person transmits a fixed, clear proposition to another by means of expressions in a common language, where both parties have all the relevant common knowledge, assumptions, values, etc. When the chips are down, meaning is negotiated: you slowly figure out what you have in common, what it is safe to talk about, how you can communicate unshared experience or create a shared vision. With enough flexibility in bending your world view and with luck and skill and charity, you may achieve some mutual understanding.

Communication theories based on the CONDUIT metaphor turn from the pathetic to the evil when they are applied indiscriminately on a large scale, say, in government surveillance or computerized files. There, what is most crucial for real understanding is almost never included, and it is assumed that the words in the file have meaning in themselves—disembodied, objective, understandable meaning. When a society lives by the CONDUIT metaphor on a large scale, misunderstanding, persecution, and much worse are the likely products.

Self-understanding

The capacity for self-understanding presupposes the capacity for mutual understanding. Common sense tells us that it's easier to understand ourselves than to understand other people. After all, we tend to think that we have direct access to our own feelings and ideas and not to anybody else's. Self-understanding seems prior to mutual understanding, and in some ways it is. But any really deep understanding of why we do what we do, feel what we feel, change as we change, and even believe what we believe, takes us beyond ourselves. Understanding of ourselves is not unlike other forms of understanding—it comes out of our constant interactions with our physical, cultural, and interpersonal environment. At a minimum, the skills required for *mutual* understanding are necessary even to approach *self*-understanding. Just as in mutual understanding we constantly search out commonalities of experience when we speak with other people, so in self-understanding we are always searching for what unifies our own diverse experiences in order to give coherence to our lives. Just as we seek out metaphors to highlight and make coherent what we have in common with someone else, so we seek out *personal* metaphors to high-light and make coherent our own pasts, our present activities, and our dreams, hopes, and goals as well. A large part of self-understanding is the search for appropriate personal metaphors that make sense of our lives. Self-understanding requires unending negotiation and renegotiation of the meaning of your experiences to yourself. In therapy, for example, much of self-understanding involves consciously recognizing previously unconscious metaphors and how we live by them. It involves the constant construction of new coherences in your life, coherences that give new meaning to old experiences. The process of self-understanding is the continual development of new life stories for yourself.

The experientialist approach to the process of self-understanding involves:

> Developing an awareness of the metaphors we live by and an awareness of where they enter into our everyday lives and where they do not
>
> Having experiences that can form the basis of alternative metaphors
>
> Developing an "experiential flexibility"
>
> Engaging in an unending process of viewing your life through new alternative metaphors

Ritual

We are constantly performing rituals, from casual rituals, like making the morning coffee by the same sequence of steps each day and watching the eleven o'clock news straight to the end (after we've seen it already at six o'clock); to going to football games, Thanksgiving dinners, and university lectures by distinguished visitors; and so on to the most solemn prescribed religious practices. All are repeated structured practices, some consciously designed in detail, some more consciously performed than others, and some emerging spontaneously. Each ritual is a repeated, coherently structured, and unified aspect of our experience. In performing them, we give structure and significance to our activities, minimizing chaos and disparity in our actions. In our terms, a ritual is one kind of experiential gestalt. It is a coherent sequence of actions, structured in terms of the natural dimensions of our experience. Religious rituals are typically metaphorical kinds of activities, which usually involve metonymies—real-world objects standing for entities in the world as defined by the conceptual system of the religion. The coherent structure of the ritual is commonly taken as paralleling some aspect of reality as it is seen through the religion.

Everyday personal rituals are also experiential gestalts consisting of sequences of actions structured along the natural dimensions of experience—a part-whole structure, stages, causal relationships, and means of accomplishing goals. Personal rituals are thus natural kinds of activities for individuals or for members of a subculture. They may or may not be metaphorical kinds of activities. For example, it is common in Los Angeles to engage in the ritual activity of driving by the homes of Hollywood stars. This is a metaphorical kind of activity based on the metonymy THE HOME STANDS FOR THE PERSON and the metaphor PHYSICAL CLOSENESS IS PERSONAL CLOSENESS. Other everyday rituals, whether metaphorical or not, provide experiential gestalts that can be the basis of metaphors, e.g., "You don't know what you're opening the door to," "Let's roll up our sleeves and get down to work," etc.

We suggest that

The metaphors we live by, whether cultural or personal, are partially preserved in ritual.

Cultural metaphors, and the values entailed by them, are propagated by ritual.

Ritual forms an indispensable part of the experiential basis for our cultural metaphorical systems. There can be no culture without ritual.

Similarly, there can be no coherent view of the self without personal ritual (typically of the casual and spontaneously emerging sort). Just as our personal metaphors are not random but form systems coherent with our personalities, so our personal rituals are not random but are coherent with our view of the world and ourselves and with our system of personal metaphors and metonymies. Our implicit and typically unconscious conceptions of ourselves and the values that we live by are perhaps most strongly reflected in the little things we do over and over, that is, in the casual rituals that have emerged spontaneously in our daily lives.

Aesthetic Experience

From the experientialist perspective, metaphor is a matter of *imaginative rationality*. It permits an understanding of one kind of experience in terms of another, creating coherences by virtue of imposing gestalts that are structured by natural dimensions of experience. New metaphors are capable of creating new understandings and, therefore, new realities. This should be obvious in the case of poetic metaphor, where language is the medium through which new conceptual metaphors are created.

But metaphor is not merely a matter of language. It is a matter of conceptual structure. And conceptual structure is not merely a matter of the intellect—it involves all the natural dimensions of our experience, including aspects of our sense experiences: color, shape, texture, sound, etc. These dimensions structure not only mundane experience but aesthetic experience as well. Each art medium picks out certain dimensions of our experience and excludes others. Artworks provide new ways of structuring our experience in terms of these natural dimensions. Works of art provide new experiential gestalts and, therefore, new coherences. From the experientialist point of view, art is, in general, a matter of imaginative rationality and a means of creating new realities.

Aesthetic experience is thus not limited to the official art world. It can occur in any aspect of our everyday lives— whenever we take note of, or create for ourselves, new coherences that are not part of our conventionalized mode of perception or thought.

Politics

Political debate typically is concerned with issues of freedom and economics. But one can be both free and economically secure while leading a totally meaningless and empty existence. We see the metaphorical concepts of FREEDOM, EQUALITY, SAFETY, ECONOMIC INDEPENDENCE, POWER, etc., as being different ways of getting *indirectly* at issues of meaningful existence. They are all necessary aspects of an adequate discussion of the issue, but, to our knowledge, no political ideology addresses the main issue head-on. In fact, many ideologies argue that matters of personal or cultural meaningfulness are secondary or to be addressed later. Any such ideology is dehumanizing.

Political and economic ideologies are framed in metaphorical terms. Like all other metaphors, political and economic metaphors can hide aspects of reality. But in the area of politics and economics, metaphors matter more, because they constrain our lives. A metaphor in a political or economic system, by virtue of what it hides, can lead to human degradation.

Consider just one example: LABOR IS A RESOURCE. Most contemporary economic theories, whether capitalist or socialist, treat labor as a natural resource or commodity, on a par with raw materials, and speak in the same terms of its cost and supply. What is hidden by the metaphor is the nature of the labor. No distinction is made between meaningful labor and dehumanizing labor. For all of the labor statistics, there is none on *meaningful* labor. When we accept the LABOR IS A RESOURCE metaphor and assume that the cost of resources defined in this way should be kept down, then cheap labor becomes a good thing, on a par with cheap oil. The exploitation of human beings through this metaphor is most obvious in countries that boast of "a virtually inexhaustible supply of cheap labor"—a neutral-sounding economic statement that hides the reality of human degradation. But virtually all major industrialized nations, whether capitalist or socialist, use the same metaphor in their economic theories and policies. The blind acceptance of the metaphor can hide degrading realities, whether meaningless blue-collar and white-collar industrial jobs in "advanced" societies or virtual slavery around the world.

Afterword

Collaborating on this book has given us the opportunity to explore our ideas not only with each other but with literally hundreds of people—students and colleagues, friends, relatives, acquaintances, even strangers at the next café table. And after having worked out all of the consequences we could think of, for philosophy and for linguistics, what stands out most in our minds are the metaphors themselves and the insights they have given us into our own daily experiences. We still react with awe when we notice ourselves and those around us living by metaphors like TIME IS MONEY, LOVE IS A JOURNEY, and PROBLEMS ARE PUZZLES. We continually find it important to realize that the way we have been brought up to perceive our world is not the only way and that it is possible to see beyond the "truths" of our culture.

But metaphors are not merely things to be seen beyond. In fact, one can see beyond them only by using other metaphors. It is as though the ability to comprehend experience through metaphor were a sense, like seeing or touching or hearing, with metaphors providing the only ways to perceive and experience much of the world. Metaphor is as much a part of our functioning as our sense of touch, and as precious.

References

Bolinger, Dwight. 1977. *Meaning and Form*. London: Longman's.

Borkin, Ann. In press. *Problems in Form and Function*. Norwood, N.J.: Ablex.

Cooper, William E., and Ross, John Robert. 1975. "World Order." In Robin E. Grossman, L. James San, and Timothy J. Vance, eds., *Functionalism*. Chicago: Chicago Linguistic Society (University of Chicago, Goodspeed Hall, 1050 East 59th Street).

Davidson, Donald. 1978. "What Metaphors Mean." *Critical Inquiry* 5:31–47.

Frege, Gottlob. 1966. "On Sense and Reference." In P. Geach and M. Black, eds., *Translation from the Philosophical Writings of Gottlob Frege*. Oxford: Blackwell.

Grice, H. P. 1957. "Meaning." *Philosophical Review* 66:377–88.

Lakoff, George. 1972. "Linguistics and Natural Logic." Pp. 545–665 in Donald Davidson and Gilbert Harman, eds., *Semantics of Natural Language*. Dordrecht: D. Reidel.

———. 1975. "Hedges: A Study in Meaning Criteria and the Logic of Fuzzy Concepts." Pp. 221–71 in Donald Hockney et al., eds., *Contemporary Research in Philosophical Logic and Linguistic Semantics*. Dordrecht: D. Reidel.

———. 1977. "Linguistic Gestalts." In *Proceedings of the Thirteenth Annual Meeting of the Chicago Linguistic Society*. Chicago: Chicago Linguistic Society.

Lewis, David. 1972. "General Semantics." Pp. 169–218 in Donald Davidson and Gilbert Harman, eds., *Semantics of Natural Language*.

Lovins, Amory B. 1977. *Soft Energy Paths*. Cambridge: Ballinger.

Montague, Richard. 1974. *Formal Philosophy*. Edited by Richmond Thomason. New Haven: Yale University Press.

Nagy, William. 1974. "Figurative Patterns and Redundancy in the Lexicon." Ph.D. dissertation, University of California at San Diego.

Reddy, Michael. 1979. "The Conduit Metaphor." In A. Ortony, ed., *Metaphor and Thought*. Cambridge, Eng.: At the University Press.

Rosch, Eleanor, 1977. "Human Categorization." In N. Warren, ed., *Advances in Cross-Cultural Psychology*, vol. 1. New York: Academic Press.